Navigator to Hydrographer

by Tom McCulloch

© Copyright 2005 Tom McCulloch.
All rights reserved. No part of this publication may be reproduced, stored in a retrieval system, or transmitted, in any form or by any means, electronic, mechanical, photocopying, recording, or otherwise, without the written prior permission of the author.

Note for Librarians: a cataloguing record for this book that includes Dewey Decimal Classification and US Library of Congress numbers is available from the Library and Archives of Canada. The complete cataloguing record can be obtained from their online database at:
www.collectionscanada.ca/amicus/index-e.html
ISBN 1-4120-4592-4
Printed in Victoria, BC, Canada

TRAFFORD

Offices in Canada, USA, Ireland, UK and Spain
This book was published *on-demand* in cooperation with Trafford Publishing. On-demand publishing is a unique process and service of making a book available for retail sale to the public taking advantage of on-demand manufacturing and Internet marketing. On-demand publishing includes promotions, retail sales, manufacturing, order fulfilment, accounting and collecting royalties on behalf of the author.

Book sales for North America and international:
Trafford Publishing, 6E–2333 Government St.,
Victoria, BC v8t 4p4 CANADA
phone 250 383 6864 (toll-free 1 888 232 4444)
fax 250 383 6804; email to orders@trafford.com

Book sales in Europe:
Trafford Publishing (UK) Ltd., Enterprise House, Wistaston Road Business Centre, Wistaston Road, Crewe, Cheshire cw2 7rp UNITED KINGDOM
phone 01270 251 396 (local rate 0845 230 9601)
facsimile 01270 254 983; orders.uk@trafford.com

Order online at:
www.trafford.com/robots/04-2400.html

10 9 8 7 6 5 4 3 2

Table of Contents

List of Illustrations iii

Foreword by Michael Bolton vii

Foreword by Rear Admiral G.S. Ritchie C.B. D.S.C. RN Ret'd ix

Preface xiii

Acknowledgements xvii

1. Becoming a Canadian 1
2. Trying to Swallow the Anchor 9
3. Sailing the Great Lakes and Other Canadian Waters 25
4. Swallowing That Blasted Anchor Again 33
5. Ashore and Afloat 43
6. Return to the Ocean 50
7. To the River Plate and Rio Grande Do Sul 61
8. Eastward across the Oceans to the Orient and the Antipodes 67
9. From Australia to Home via the Panama Canal 74
10. Hydrographic Surveying the Waters of British Columbia 80
11. Enjoying the Local Scene – The Year of the Miracle Mile 91
12. Armed with My Master's Foreign-Going – Another Hydrographic Season 101
13. More of the Same – But Changes in the Offing for 1957 106
14. New Responsibilities – Ganges Harbour and the Western Arctic 114
15. The Ottawa Scene – Fulford Harbour – *Storis* Again – A Sad Ending to the Year 124
16. A Canadian Presence in the Western Arctic 131

17.	A Dubious Venture – A Lesson Learned	137
18.	CSS *Richardson*'s Voyage into the Western Arctic—1962	145
19.	1963 – A Good Ice Year	157
20.	1964 – A Very Bad Ice Year – Family Transported to Ottawa	160
21.	Exposure to the Wave of the Future – A Blow to the CHS Ego	167
22.	1967 Voyage to the Western Arctic – An Epic Venture	173
23.	Changes in Direction – Becoming a Desk Jockey	184
24.	Regional Responsibilities	191
25.	A New Base – The CCIW Fleet – Searching for Pingoes	198
26.	Expanding Horizons – R&D and Europe	207
27.	Changes in the Management Structure – Becoming a Bureaucrat	217
28.	Weathering the Bureaucracy – Broadening My Horizons	222
29.	Pushing the Profession	229
30.	Fifty Years – How Could That Be?	237
31.	The Pace Quickens	249
32.	Actions at Home and Abroad	255
33.	Major Changes in the Hierarchy	266
34.	Triumphs – Rewards – Reflections	270

List of Illustrations

Figure 1 - *Wm. J. Stewart*, Grenville Channel 1954

Figure 2 - When we had a scientific fleet on the Pacific Coast, 1970 - *Baffin*, *Hudson*, *Parizeau*, alongside wharf, *Quadra* in dry dock

Figure 3 - Port Arthur CPR station, 1948...3

Figure 4 - The Great Lakes of North America 1948....................................4

Figure 5 - Port Arthur harbour, 1940s..9

Figure 6 - Around the Globe voyage of *Seaside* Part I, 1948–49....................15

Figure 7 - Around the Globe voyage of *Seaside* Part II, 1948–49...................16

Figure 8 - SS *Shelton Weed*, 1949..27

Figure 9 - SS *Royalton*, 1949..29

Figure 10 - SS *Vandoc*, 1951...45

Figure 11 - The Great Lakes 47

Figure 12 - SS *Angusglen*, 1951/1952..50

Figure 13 - Stan Hollis, VC, Junior Engineer, with radio officer, *Angusglen*, 1951/1952 54

Figure 14 - My own second mate, *Angusglen*, 1951/1952............................56

Figure 15 - *Angusglen* voyage around the Globe, Part I, 1951/52...................62

Figure 16 - My usual working rig..63

Figure 17 - *Angusglen* awaiting rice at Rio Grande do Sul, 1952 65

Figure 18 - *Angusglen* voyage around the Globe Part II, 1951/52...................69

Figure 19 - Entering Port Phillip Bay preparatory to loading cargo in the Port of Melbourne, July 1952..74

Figure 20 - The Pacific Coast of Canada 81

Figure 21 - CSS *Wm. J. Stewart*, 1953 82

Figure 22 - The Clash of '53 84

Figure 23 - Our double-ended "covered wagons," 1953...............................86

Figure 24 - The salvage of the *Wm. J. Stewart*, 1944 89

Figure 25 - Launches ready for action, 1954 94
Figure 26 - Operational area in the Queen Charlottes, 1954 95
Figure 27 - Operational area, Queen Charlottes, 1955 102
Figure 28 - Poole, Sandilands, Wills, with Al Ages in front, Butedale 1955 103
Figure 29 - Crew of *Port 2*, Queen Charlotte Islands 1955 104
Figure 30 - Operational area, Queen Charlottes 1956/57 108
Figure 31 - Decca 6f trials and tribulations, 1957 111
Figure 32 - Our young eagle "Joey," 1957 111
Figure 33 - Ripple Rock in Seymour Narrows 116
Figure 34 - Explosion at Ripple Rock in 1958 116
Figure 35 - USCGC *Storis*, 1958 117
Figure 36 - Voyage to the Arctic, 1958 119
Figure 37 - Seasonal route through the Western Arctic 120
Figure 38 - CCGC *Camsell*, 1960 131
Figure 39 - RCMP launch *Spalding*, Cambridge Bay, 1961 140
Figure 40 - CSS *Richardson* awaiting launch, Star Shipyard, Annacis Island, March 1962 145
Figure 41 - Dignitaries present at *Richardson* launch, March 1962 146
Figure 42 - Mavis Young successfully christens *Richardson*, 1962 147
Figure 43 - Sir John Richardson 147
Figure 44 - The long voyage in 1962 150
Figure 45 - CSS *Richardson* sheltering in the ice pack, Beaufort Sea 1962 152
Figure 46 - Ron Longbottom deploying the hand lead, shallow waters off the Mackenzie Delta, 1962 153
Figure 47 - Sounding operations in the Beaufort Sea, author and Ron Card, 1963 159
Figure 48 - In heavy ice concentration, 1964 162
Figure 49 - My gang! 1965 – baby Sarah at home 166
Figure 50 - Surveying from the upper bridge of the *Wm. J. Stewart*, 1966 169
Figure 51 - The 4047 nautical miles between Victoria and Tuktoyaktuk 174
Figure 52 - CSS *Richardson* in trouble – vicinity of Point Barrow, July 1967 176

Figure 53 - Captain Strand from *Camsell* comes calling, July 1967 177

Figure 54 - *Northwind* to the rescue, July 1967 178

Figure 55 - The Arctic train: *Northwind*, *Camsell*, taken from little *Richardson*, July 1967 179

Figure 56 - The long tow along the north coast of Alaska, July / August 1967 .. 180

Figure 57 - The news from home, 1967 181

Figure 58 - A family welcome at Victoria Airport, September 1967 182

Figure 59 - Sarah's reluctant welcome, 1967 183

Figure 60 - Canada Centre for Inland Waters, Burlington Ontario, 1970 198

Figure 61 - CSS *Limnos*, Great Lakes 1974 200

Figure 62 - CSS *Parizeau*, Pacific Coast 1970 202

Figure 63 - CSS *Port Dauphine*, Welland Canal 1970 205

Figure 64 - CSS *Bayfield*, Burlington Bay 1970 208

Figure 65 - Rear Admiral G.S. Ritchie RN CB DSC, Chairman, Hydrographic Commission, FIG, Wiesbaden 1971 212

Figure 66 - CSS *Advent*, fast cutter, Burlington 1972 220

Figure 67 - With my brother Gordon on his Torkington estate, Malaysia 1973 226

Figure 68 - CIS: having fun in Fredericton, June 1974 231

Figure 69 - Fifty and suddenly aware of it! 1975 237

Figure 70 - Left to Right: Rear Adm. David Haslam RN, Mike Bolton, Regional Hydrographer Pacific, accepting "Fickle Finger of Fate" from Adam Kerr, Regional Hydrographer Central, 1977 259

Figure 71 - Willie Rapatz gives Lighthouse award to Sandy Sandilands, 1977 .. 259

Figure 72 - A brace of Admirals (Ritchie and Haslam), 1977 260

Figure 73 - The plaque commemorating the first IHTC, Ottawa 1979, sponsored by Fédération Internationale des Géomètres, Canadian Hydrographic Service, Canadian Hydrographers Association, and the Canadian Institute of Surveying 270

Figure 74 - Rear Admiral Henry Wolsey Bayfield RN, 1795–1885 274

Foreword by Michael Bolton

This is Volume II in the life story of Tom McCulloch. It covers the period from when he and his wife Doreen arrived in Canada in 1948 until 1979, when he was firmly established in Burlington. During that time frame many changes occurred in both his personal and public life. He became the devoted father of five children and rose to a senior position within the Canadian Hydrographic Service.

When I originally read the book, I suggested that Tom consider an alternative title such as "From the Private Sector to Public Service," but he would not hear of it. Hey, that's OK, he's the author—therefore he's the boss.

A little background on hydrography in Canada may be helpful in the context of the story. After WW II, Canadian hydrography, particularly within the Government of Canada, grew at an amazing rate. There seemed to be loads of money, and new positions became available to recruit additional staff. Personnel were hired either with maritime backgrounds or surveying education. Consequently, merchant marine officers, both Canadian and British, graduates from universities in engineering, as well as institute of technology graduates in surveying were all actively recruited.

These new staff, with varying backgrounds, were originally trained on the job, but it was soon realized that this was not good enough. Formal training programs were created and new personnel were exposed to these programs. Eventually a ship was dedicated to the task, so the staff could receive both classroom and field experience. In later years, formal hydrographic training was recognized internationally, with programs being established around the world.

In addition to new staff, new equipment was procured as technological changes were altering the face of hydrography. The visual "fix," the positioning of a ship or launch by visual methods, usually a sextant, was gradually being replaced by electronic systems that reduced the "hands-on" requirements. Echo sounders were becoming more sophisticated so that total ocean bottom coverage could be obtained through the use of side scan sonar. Other changes also occurred, but with lots of funding available, we were able to stay abreast of all the latest systems.

Not only staff and equipment were expanded and enhanced. In 1950, within the Canadian Hydrographic Service, there were headquarters in Ottawa and the Pacific Region in Victoria, BC. By 1975, CHS had three regions in addition to HQ. The Atlantic Region, located in Dartmouth, NS, was housed in the well-established Bedford Institute of Oceanography. Central Region by now called Burlington, Ontario, home and resided in the accommodating Canada Centre for Inland Waters, while Pacific Region relocated from its downtown shabby offices to the magnificent Institute of Ocean Sciences at Patricia Bay on the Saanich Inlet, 20 miles north of Victoria.

Along with new facilities came new ships and launches, equipped to handle the latest techniques in the acquisition of hydrographic and marine scientific data. Some of the vessels were designed to operate in ice-infested waters, specifically the Canadian Arctic. These craft were used extensively throughout the time frame.

External relationships flourished. Through the Canadian Institute of Surveying (Tom is a past president), communication with national and provincial survey and mapping agencies were maintained. The International Hydrographic Organization, based in Monaco, provided entrance to government hydrographic establishments throughout the world. International non-governmental access was obtained through the Fédération Internationale des Géomètres. Communication with our American counterparts was maintained through various working groups between ourselves and our National Oceanic and Atmospheric Administration allies. All in all, we had excellent national and international rapport proceeding.

Despite all the excellence described above, we did have our internal squabbles and disagreements. Regions were always demanding a larger share of the pie, be it man-years or money or ships. Each Region evolved slightly differently, so there were often debates as to who was correct, when many times the right answer was in the eye of the beholder. But through all these minor disputes, Canadian hydrography during the time in question could be considered to be healthy, wealthy, and wise.

So with this brief background I invite you to peruse Volume II of the ongoing saga of T.D.W. McCulloch.

        Michael Bolton
        Regional Director of Hydrography, Pacific Region, Ret'd
        Long time senior and influential official of the Canadian
        Hydrographic Service

Foreword by Rear Admiral G.S. Ritchie C.B. D.S.C. RN Ret'd

The last photograph in Tom McCulloch's book *Mandalay to Norseman* shows a handsome young Scottish couple about to emigrate to Canada in 1948.

In the early chapters of this book Tom relates how they settled into their new country, which in time they came to love.

Tom is first and foremost a seaman who has sailed on many a ship, and for the first five years in Canada he earned his livelihood as a deck officer in merchant ships trading both world wide and in the Great Lakes, leaving his wife Doreen for many months at a time at their first home at Port Arthur at the head of Lake Superior.

Tom never introduces a ship without giving her full particulars—tonnage, length, draught, speed etc., and presents thumbnail sketches of his shipmates, sometimes relating their experiences on hilarious runs ashore many years ago.

During his voyaging he always hoped that one day he could swallow the anchor, so when he saw an advertisement seeking qualified seamen to join the Canadian Hydrographic Service, he applied. He was accepted at the interview as a technical employee and was soon despatched to the west coast to join the renowned survey ship *Wm. J. Stewart* surveying in British Columbian waters.

Canadian Hydrographic ships are conducted by a Master and his Mates to comply with the surveying requirements of the hydrographer-in-charge onboard. So although Tom qualified as a Foreign-Going Master, he had now crossed the Rubicon to join other qualified mariners who had become hydrographers.

As a junior surveyor he spent the long summer seasons in the ship's sounding launches for several years before he was sent to the Western Arctic, where the United States government was establishing the DEW Line stations along the Canadian Arctic shoreline.

After three Arctic seasons serving in US and Canadian ships which were not suited for surveying the approaches to the DEW Line stations, McCulloch, on his annual visits to Ottawa, convinced the authorities that a dedicated vessel should be built on the west coast and made available for the Arctic survey tasks.

And so it came about that a steel-hulled vessel about sixty-six feet in length, to be named after Dr. Richardson, an Arctic explorer of the early nineteenth century, was built at New Westminster under McCulloch's general supervision. He took command in 1962 and after extensive sea trials sailed for the Arctic. It must have been a wonderful day for Tom when he passed Point Barrow and sailed eastward. He was to achieve four useful summer seasons surveying with modern fixing equipment such as Decca, tellurometers, and radar ranging.

In 1967 a number of surveyors, including Tom, were assembled in Ottawa to be considered for advancement to senior management posts. Dr. Cameron, Director of the Marine Science Department, the chief interviewer, bluntly informed them that their wider knowledge of marine science was generally inadequate.

These unhappy hydrographers travelled home dejectedly. However, they slowly got together over the next year or two to form the Canadian Hydrographic Association to put pressure on Management to promote professionalism with the introduction of a formal system of training and education of their surveyors.

Tom became the second president of the CHA, but not for long, for at the conclusion of a CHS Conference in Ottawa in 1968, to Tom's amazement, Dr. Cameron announced that he would be appointed a Regional Hydrographer. So he had again crossed a divide to join the Management.

He took over the administration of the Central Region, which covered the Great Lakes, the lower St. Lawrence River, and the Arctic, with his office in Ottawa. This meant a move of the family from Victoria.

There was a Canada Centre for Inland Waters at Burlington on Lake Ontario with ships, launches, and workshops, and that is where Tom considered his Regional Office should be. He used his gift of persuasion, not for the first time, and Dr. Cameron approved the idea. Doreen's move to Burlington took place in 1970.

McCulloch stayed there until his retirement, by which time the Burlington establishment had been renamed the Bayfield Laboratory of Marine Sciences and Surveys with Tom as Director General.

There is a heroine in this long story—Doreen, who had brought up a family of five children with her husband so often far away for long periods of time.

As current president of the Canadian Institute of Surveying, Tom was expecting to give the dinner speech at the Halifax Conference, but as it was Woman's Year, the Committee had engaged Doreen for the task. Tom has been gracious in reproducing her speech word for word in the book. Every surveying wife should be given a chance to read it; I know they will appreciate Doreen's carefully aimed arrows.

Tom's interests were ever widening, and in the later chapters he is travelling the world, first in the interests of the Canadian Hydrographic Service, and then as a member of the Fédération Internationale des Géomètres (FIG) and the Commonwealth Association of Surveying and Land Economy (CASLE). He takes us with him, and how engaging it is, for he describes in critical detail every city he visits, every hotel he stays in, from the very best to the very worst. He meets many characters along the way, each of which he brings to life, and he certainly has an eye for beautiful women!

In the final chapter McCulloch summarises his tale – 'from being an almost penniless emigrant from Britain to a comfortable bureaucrat.' "Formidable" would be a more suitable description of this bureaucrat, whilst this book is a major contribution to the history of Canadian hydrography.

Rear Admiral G.S. Ritchie C.B. D.S.C. RN Ret'd
Former Hydrographer of the Royal Navy
and President of the Directing Committee of the
International Hydrographic Organization

Preface

When writing the preface to my previous book *Mandalay to Norseman*, I noted particularly the encouragement that I received from my wife and children as a spur to eventually putting pen to paper. This time around, although the family are still very much in my corner, I now very much want to record my view of what transpired after arriving in Halifax, Nova Scotia, as a landed immigrant to Canada in June 1948.

The Canadian Lakehead became my base of operations as I struggled to settle in Canada while continuing in an on and off again manner my chosen profession of the sea. Attempts to swallow the anchor were followed by voyages around the world and by traveling the inland waters of the St. Lawrence River and the Great Lakes of North America.

Then, in 1953, I made the move that was to affect my life significantly and set me off on a different course—the sea would still be part of my life but in a different way. I joined the Canadian Hydrographic Service as a Hydrographic Surveyor in training and thereby became a collector and depicter of nautical data that I had formerly relied upon as a navigator. What follows is not a history of the Canadian Hydrographic Service but my very own personal recollection of events and people who were part of my experiences.

My career with the Canadian Hydrographic Service commenced in late May 1953 when I joined CSS *Wm. J. Stewart* on the Pacific Coast of Canada. In the book I describe life and work as a hydrographic surveyor on board a survey vessel engaged in hydrographic activities ranging from Prince Rupert and the Queen Charlotte Islands to Johnstone Strait and the Broughton Archipelago, together with the first electronically controlled surveys of Hecate Strait. In more southern waters we carried out revisory surveys of major harbours such as Vancouver, Victoria, and Nanaimo.

In 1958 I was given as my first command surveys of harbours on Saltspring Island, while commencing an association with the Western Arctic as hydrographer-in-charge on US and Canadian vessels that lasted until 1967 when I commanded CSS *Richardson* and came close to losing her and our crew in sea ice under pressure off Point Barrow, Alaska.

The middle sixties had also seen my growing fixation with the urgent need for education and training of hydrographers to keep abreast of the wave of new technology and technique becoming available to the profession in government service and in the expanding private sector.

In 1968 I found myself behind a desk directing the activities of the Central region of the Canadian Hydrographic Service, at that time based in Ottawa, but located separately from CHS Headquarters. After two years the region moved lock, stock, and barrel out of Ottawa to the Canada Centre for Inland Waters at Burlington to join our fleet of vessels and machine shops already located there.

Years later I became Director Marine Sciences and Surveys for the region, and although still retaining operational control of hydrographic activities in the region, was no longer an integral part of CHS management where policy and long-term planning were decided. My main activities shifted more to international affairs, such as the International Field Year Great Lakes, the Fédération Internationale des Géomètres, the Commonwealth Association of Land Economy, and to the surveying profession within Canada, together with taking part in national marine sciences management decision making on policy matters and the development of a coherent strategy for our enormous exclusive economic zone. An additional responsibility was the regional build-up of an oceanographic competence that would concentrate its efforts in the High Arctic utilizing the support provided by the Polar Continental Shelf Project (PCSP).

My work in the profession brought me the honour of becoming, in 1974, the first president of the Canadian Institute of Surveying with a hydrographic surveying background, and in 1977, vice chairman of FIG Hydrographic Commission and a member of the executive of CASLE. Finally, with much help from all these organizations and in particular the management of the Canadian Hydrographic Service, we were able to host the very first International Hydrographic Technical Conference (IHTC). It was held in Ottawa in May 1979 and was a great success both professionally and technically. I was much pleased!

During those years between 1973 and 1979, I became a world traveler and revelled in the opportunity to meet new acquaintances and old friends in various parts of the globe. It was exciting but sometimes tiring. In 1979 I was confirmed Director General of the Bayfield Laboratory for Marine Science and Surveys, which is a useful marker in time. My tale therefore ends in 1979.

If health and strength and the will of the Almighty prevail, a further volume could be forthcoming, going on from 1979 to more recent events that might interest the reader.

I sincerely thank all those shipmates, surveyors, and others who helped make my career interesting and productive, and I believe worth recording. We lived through the very best years of the Canadian Hydrographic Service.

Acknowledgements

This story—*Navigator to Hydrographer*—is largely drawn from my memory of meaningful events that I can recall from my first thirty-one years in Canada. I also gratefully acknowledge the help of a number of people who were able to corroborate some of my recollections of times past or were able to correct some of my more fanciful assertions.

For the *Wm. J. Stewart* period, 1953 to 1957, I am particularly grateful for the input provided by Messrs Sandy Sandilands, Al Ages, and Willie Rapatz, who still remembered many of the incidents described.

For the Western Arctic days, 1958 to 1967, old shipmates such as Bob Weinberg, Al Ages, Willie Rapatz, Ron Longbottom, Stan Huggett, and Barry Lusk provide much useful insight, while Tony Mortimer, Alan Meadows, and Art Mountain were actual witnesses when *Richardson* almost foundered in the ice under pressure off Cape Barrow, Alaska. I am indebted to Pamela Olsen, Librarian at the Institute of Ocean Sciences, Patricia Bay, BC, for her assistance in obtaining access to the many relevant reports and studies covering these activities in the Western Arctic, and to Sandy Sandilands, whose book *The Chartmakers* was an invaluable source of hydrographic information.

For other events from the mid sixties to 1979 I owe much to Barrie MacDonald, Mike Bolton, and Steve MacPhee, who tracked the facts for me and made sure that I made no egregious error.

To all those who provided me with suitable illustrations and to any sources of material that I may have inadvertently overlooked, I gratefully acknowledge your assistance.

Mike Bolton, in particular, I sincerely thank for our continuing association and friendship over many years, and for his kind consent to dedicate a Canadian foreword to *Navigator to Hydrographer*. He was my constant link to the Canadian Hydrographic Service in times of stress and in times of achievement.

I wish to extend my gratitude to Rear Admiral Steve Ritchie CB DSC, former Hydrographer of the Royal Navy and President of the Directing Committee of the International Hydrographic Organization, a good friend and colleague for many years, who imparted much good advice from time to time and kindly consented to dedicate an international foreword to my publication.

Finally, Claire Champod's interest and dedication to editing, proof reading, and general layout of the book for printing are much appreciated, as are Gail Taylor's efforts in the final few months.

Figure 1 - *Wm. J. Stewart*, Grenville Channel 1954

Figure 2 - When we had a scientific fleet on the Pacific Coast, 1970 - *Baffin*, *Hudson*, *Parizeau*, alongside wharf, *Quadra* in dry dock

Chapter One

Becoming a Canadian

In *Mandalay to Norseman*, I concluded the book by describing my wife and me devouring our first banana split upon arrival in Halifax from Liverpool. It was June 1948 and we had been accepted as brand new immigrants to Canada.

That, evening our train departed for Montreal, a full day's journey away. The rattle of the rails and the swaying of the carriage soon lulled us off to sleep in our curtained bunks. I think that I lost the draw and ended up in the top bunk.

By early the next morning, we had moved out of the English-speaking areas and were deep in the French-speaking areas of the Gaspé Peninsula. By breakfast in the dining car, we were running along parallel to the St. Lawrence River and heading in a southwesterly direction toward Montreal. Our train was Canadian National out of Halifax. We learned that Canadian Pacific commenced in Saint John, New Brunswick, and crossed through the state of Maine before entering the province of Quebec much closer to Montreal. We did not resent the extra time spent on our journey up the lovely St. Lawrence River valley. It was different from my memories of shipping on the river itself, but just as awe-inspiring. Canada truly was a mighty large country.

Late that afternoon, our train pulled into Central Station in Montreal, where Doreen's relatives welcomed us. The train had been packed with other immigrants who all seemed to vanish as soon as the train disembarked. We were left facing people we had never seen before, except in faded photographs taken back in England many years before. However, it was not too long before we had sorted out who was who among all those greeting us. There was Aunt Agnes, sister of Doreen's father, and her husband Clem, and their daughter Edna. Then, to represent Doreen's mother's family was her uncle Charlie, his wife Bella, and their son Stewart. It was a happy occasion for all of us, even if somewhat confusing for the new arrivals.

We were to stay with Aunt Agnes at first, then move over to Uncle Charlie's later. Aunt Agnes and her family lived in NDG (Notre Dame de Grâce), in a large, rather pleasant first-floor apartment. It appeared to be a largely English-speaking area, with lots of small parks, full of gardens and deciduous trees. After a settling-in period with family, we set out to explore Montreal.

In 1948, Montreal was a large bustling industrial city and seaport, with a population in excess of two million souls, of whom seventy-five percent were of French origin and the remainder largely English speaking. It was the largest city in Canada, and the centre of power in commerce and industry. It tended to regard Toronto as a stiff, dull, smaller city that was in no way a worthy competitor. Montreal was vibrant with a strong Gallic flavour diffused with English power and know-how. Its main east-west artery, rue Ste-Catherine, was an indicator of the racial mixture of the city, with French predominant in the east end and English in the west end. An enormous variety of clothing shops were situated in the core section of St. Catherine, ranging from the very elegant and expensive to the large department stores and the small inexpensive shops specializing in smaller selections of goods. The other streets around the St. Catherine core were filled with tall buildings housing many of Canada's most prominent companies' head offices.

Leaving my better half to view and partake of the goodies displayed on St. Catherine, I set off to find myself employment by visiting likely companies located in the centre of the city. Within a short time I had several offers of employment on Canadian merchant ships, but settled with Saguenay Terminals where I agreed to report for duty in early July, after returning from Port Arthur where I would leave my wife with her family. A second mate's berth was offered at three hundred dollars a month, which seemed reasonable to me and more than generous according to my wife's aunt. A celebration was in order and we duly complied by visiting several bars and nightclubs to soak up the local atmosphere.

In the days following, we did the tourist thing, visiting points of interest such as the Cross on Mount Royal, the Catholic Cathedral, and the Old Port of Montreal. It was fascinating, but we were now looking forward to our journey westward. Toward the end of June we made our farewells and joined a Canadian Pacific train at Windsor Station. Soon, we were rattling our way across northern Ontario, past the desolation of the mines around Sudbury, past a myriad of small lakes, before finding ourselves running along the north shore of Lake Superior, in the vicinity of Marathon. The train was comfortable, the bunks narrow but well padded, and the food served in the dining car of good quality. We were once again appreciating the immense size of Canada. After twenty-six hours steaming from Montreal we pulled into Port Arthur early in the evening, to be greeted by Doreen's family and many others. We were a long way from home!

Figure 3 - Port Arthur CPR station, 1948

Port Arthur, Ontario, was one of two small cities located at the Canadian head of the Great Lakes system. It lay on Lake Superior, just to the east of Fort William, which is located on the River Kam. (By 1970, these would be amalgamated to form Thunder Bay.) In 1948, each city had a population of around thirty thousand, with very mixed ethnic backgrounds of English, Scots, Finns, Ukrainians, Italians, Yugoslavs, Germans, Swedes and the occasional French Canadian—a polyglot crew indeed. They worked in the surrounding forests, the paper mills, the grain elevators, the shipyard, and in the manufacturing of vehicles and military aircraft. The results of their labour were shipped by lake from the many wharves and jetties along the waterfronts of both cities. There was a fairly friendly rivalry between the cities, reflected in the winter by tussles for dominance in ice hockey. Port Arthur also prided itself on its attractive scenery rising from the waterfront, while Fort William made do with the flat plain through which flowed the River Kam. This was to be our main base of operations for a number of years, as we would gradually realize. It was a long way from southern Ontario, where most British immigrants settled, with vastly more opportunities for compatible work. That fact would become more significant as the years went by. My sister-in-law Ursula should have married a Canadian airman from Toronto instead of her ex-flyer from the Lakehead.

None of the foregoing was on our minds as we were warmly greeted at the Port Arthur railroad station. We were borne off to my sister-in-law's residence, where we fitted comfortably into an already full house containing not only Ursula, her husband, Bobbie Guerard, and son Anthony, but also Doreen's other sister, Joan, who had come out from England in 1947, plus her mum and dad who had arrived in Canada one month before us. It was a packed house! After a few days getting to know people and our different surroundings, I left my wife and headed back to Montreal.

I thought that I had a fair amount of knowledge and understanding about Canada before arriving as an immigrant. After all, I had visited Quebec City and Montreal in 1945, sailed through the Gut of Canso in the Maritimes, plied the waters of the Gulf of St. Lawrence, and traversed the Straits of Belle Isle. Additionally, I had met many Canadian servicemen during the war years and thought I had soaked up a lot of information about my newly adopted land. I now know that I had much to learn.

Figure 4 - The Great Lakes of North America 1948

Canada was part of the British Empire, soon to be the British Commonwealth, and later just the Commonwealth. It was a democracy with a parliamentary system of government similar to that of the United Kingdom. It had, however, one important difference, the language and racial split between the English and French speaking parts of the country. While English predominated in the Maritime Provinces, French took over in northern New Brunswick and, of course, in the province of Quebec, its heartland. In Ontario, pockets of French gave way to English, particularly in the south and the west. Beyond Ontario the Western provinces of Manitoba, Saskatchewan, Alberta, and British Columbia were now all firmly in the English-speaking camp, although significant clusters of French speakers could be found in Manitoba and Saskatchewan, and to a lesser extent in Alberta. The total population was about thirteen million souls in 1948, with most of us living clustered close to the border with the USA. Although we traded a great deal with our southern neighbour, much of our trade in both manufactured goods and in raw materials was with a recovering Europe and the rest of the world.

A moderately sized Canadian-flagged merchant fleet carried our exports of automobiles, locomotives, and aircraft to the developing world, while vast quantities of wheat, lumber, and minerals went to the developed nations of the globe. The level of activity in the main seaports, such as Montreal, Vancouver, and Halifax, was a living testament to our world-recognized status as a major trader. In political terms, the Dominion Government was based in Ottawa and was responsible for national affairs, while all ten provinces conducted their matters entirely under their own jurisdiction. On paper, the boundaries between national and provincial concerns seemed reasonably clear, but in fact they were ill defined and open to dispute. This was particularly the case in matters affecting Canada as a whole and the province of Quebec. Even a newcomer like me could foresee future problems.

My return to Montreal was brief, but I was able to see a bit more of the city. It truly was a cosmopolitan area, which was established in 1642 by French settlers, who founded the village of Ville-Marie on the banks of the St. Lawrence River. It became a fortified town in 1685 and adopted the name Montreal to reflect its situation beneath the slopes of Mount Royal. In 1760, it fell to the British, and by the early nineteenth century had become the political and commercial hub of Lower Canada. It became a Victorian showcase in the later nineteenth century, as it absorbed the changes brought about by the export of the Industrial Revolution to Canada. In the twentieth century it became a metropolis and the centre of financial power in Canada. It retained that pre-eminence in 1948 and continued so until the 1960s, when commercial and financial power began to flow to its rival, the city of Toronto, capital of Ontario and the heartland of Canadian industry.

I travelled by train from Montreal to Port Alfred on the Saguenay River, a small port devoted to the discharging of bauxite ore from the West Indies for conversion into aluminium at the Alcan plants located close to the river. The air was filled with the choking dust of bauxite hanging over the wharves and the discharging vessels. It was not exactly as I would have planned. However, I duly reported on board the SS *Peribonka* as second mate. She was a nondescript tramp steamer of seven thousand tons gross with a mixed crew of Canadians and British and had been on the Port Alfred–West Indies run for quite some time.

We sailed from Port Alfred in early July 1948 bound for Georgetown, British Guinea. We were in ballast with our midship deep tanks filled with fresh water. The weather was excellent during the entire voyage, with the only hold-up an engine breakdown in the Caribbean. We wallowed in the ocean swell for several days while the ship's engineers slaved away in the intense heat of the tropics to get us mobile once more. I can still see the chief engineer, filthy with oil and sweat, raising his arms in triumph as the problem was solved and shouting up at the bridge where the Master stood: " You deck bastards are helpless without us!" A truly defining moment.

Peribonka reached Georgetown two days later. It was a hot sticky port that was also the capital of the colony. The population appeared to be evenly split between those of Indian background and those of African background. The former were descended from the indentured labourers brought into the colony from India to work in the sugar cane industry, the latter from slaves shipped in from West Africa to help in opening up the colony. There was a certain tension apparent between these two racial groups, reflected in the local political parties striving for power.

That tension was not confined to the political scene, as I soon discovered. I was instructed to report on board another Saguenay Terminals vessel already tied up to the wharf. She was a smaller vessel—4700 gross tonnage—being used to load bauxite ore up the shallow river at Mackenzie and transfer her cargo to Chaguaramas, Trinidad, for onward shipment to Port Alfred. This was an unpleasant shock to me, particularly when I saw her for the first time.

She was in light draft mode but covered from stem to stern in bauxite ore dust mixed in with residue of a tropical downpour. The ship had a dirty, uncared-for appearance and was manned by a sullen ship's company. It was a most appalling sight and a short meeting with the captain, a taciturn and scowling Geordie, did little to cheer me up. I truly had come down in the world. Paddy Henderson and Cable & Wireless seemed a distant wonderful dream.

Upon reflection, I should have refused the posting, but instead made matters worse by imbibing too much local rum and getting into a heated discussion with the captain. I packed my bags (barely unzipped) and moved ashore.

I stayed at the Anglican Mission to Seamen for several days, fighting off mosquitoes and other flying objects while alternating between feeling sorry for myself and recognizing that I had behaved like an idiot. I stayed away from the local plonk—Demerara rum —and wished I were back at sea. To my great good fortune, a berth was found for me on *Peribonka,* which had returned from a short voyage to Trinidad. We sailed from Georgetown bound for Caribbean ports. My memories of that episode in British Guinea lingered with me for many a day.

Our first stop was the bauxite ore docks in Chaguaramas, Trinidad, and then on to St. Thomas in the US Virgin Islands to top up cargo. The pier at Chaguaramas was a long way from Port of Spain, so we stayed on board, but in St. Thomas we were able to take a look around and found it to be quite an attractive spot. American beer was available, the coldest beer I ever tasted.

Peribonka had an uneventful voyage back to Port Alfred in Quebec with a full load of bauxite ore. On the trip, I was introduced to poker and lost more money than I could afford. I also spent a bit of time yarning with two of the junior engineer officers, who were Anglo-Burman in background and had vivid memories of the last days before Rangoon fell to the Japanese. According to them, the Royal Air Force's defence of the city was a washout and only the Flying Tigers put up a good fight. Was this true, or was it post-colonial bitterness?

When the ship anchored off Port Alfred, we were boarded by a large group of Customs rummagers who searched the ship from bow to stern, finding more than one hundred and fifty bottles of rum hidden in the cargo holds and various other parts of the ship. A large collective fine was placed on the ship's company, which caused much moaning and groaning. I decided to leave the ship and head directly for Port Arthur, and my wife. I needed to recuperate and recharge my batteries after bruising my ego badly in Georgetown. One last comment on that abortive voyage: In Port Alfred, before leaving by train for Montreal, I visited my first illegal drinking establishment—blind pig or whatever you like to call it. It was located in what appeared to be the root cellar of an old farmhouse. The clientele were all seafarers like me, and some of them spoiling for a fight. One French Canadian hated me on sight, as I was obviously a British import. A short tussle ensued before we were both unceremoniously ejected from the premises. I was beginning to feel that liquor and I did not mix well—I needed a shoulder to cry on.

After travelling by train to Montreal, I briefly visited with my wife's relatives in Notre Dame de Grâce, before embarking on the long train journey out to northwestern Ontario. It seemed as if it would never end—trees, rocks, rivers, small lakes, broken by the occasional small settlement out in the middle of nowhere. Then, suddenly, we were steaming along the shore of Lake Superior and, after the long haul from Montreal, we were puffing to a stop at the Canadian Pacific Railway Port Arthur Station, formerly known as Prince Arthur's Landing. It was a long way from the Caribbean and the big cities of central Canada, but here was my wife and here I would make my base of operations for the next several years, while trying to put down permanent roots.

Chapter Two

Trying to Swallow the Anchor

In 1948, the twin cities of Port Arthur and Fort William at the Canadian Lakehead had a combined population of about sixty thousand souls, split almost evenly between the two cities. The few hamlets in the cleared lands close to the cities probably supported an additional two thousand persons. As noted earlier, the area was a grand mixture of races and languages, with Ukrainian competing with Finnish and Italian to supplant the English/Scottish rump of the population that still held most of the strings of power. Additional elements were the French and, of course, the local native tribes.

This polyglot group was engaged in many activities, mostly associated with forest products and the transportation of these products—as well as grain and ore from the West—through the waterfront docks for trans-shipment by Great Lakes shipping to the deep-sea port of Montreal and beyond.

Figure 5 - Port Arthur harbour, 1940s

The forest products industry alone supported many major extraction activities north of the cities, particularly in the vicinity of Lake Nipigon, but also provided the raw material to enable four large paper mills to ship large quantities of newsprint and fine papers to other parts of North America. The grain was transported by rail from the Prairies to the Lakehead, and into grain elevators where large lake vessels loaded their cargoes for onward transportation eastward through the Great Lakes system of lakes, rivers, and locks. Iron ore was transported into the Lakehead by rail from the newly opened mines at Atikokan in northwestern Ontario.

Additionally, the cities boasted of two manufacturing plants. One, the Port Arthur Shipyard, built ships mostly for the Great Lakes trade, but also smaller vessels for shipment to Europe as part of the Marshall Plan. (The small tug that I sailed on in Egyptian waters in 1943, referred to in my book *Mandalay to Norseman*, was built in the Port Arthur Shipyard and shipped by wartime freighter around the Cape of Good Hope to Suez.) The second, the Fort William-based Canadian Car & Foundry manufactured buses and exported them all over North America.

The countryside around the Lakehead was not unattractive, with magnificent views of the seaward approaches to Port Arthur being easily obtained on the higher ground behind the city. One land mass stuck out offshore, portraying the shape of a "Sleeping Giant," named thusly by local First Nations countless centuries before the first white man appeared on the scene. Although the land around Fort William is mainly fairly flat, grand views can be had from the summit of Mount McKay, above the River Kam and, of course, in the vicinity of Kakabeka Falls. Vast evergreen forests stretched northward, seemingly forever.

The history of the Canadian Lakehead, after the arrival of the first voyageurs in the early eighteenth century seeking furs in trade with the natives, is one of gradual expansion, enhanced by political considerations that were taken in response to events taking place many miles away. The most important of those was the Riel Rebellion of 1869–70 in Manitoba, which brought about the need for a military expedition from Ontario to the Red River country to re-establish Canada's claim to the North West Territories. The United States government would not allow the expedition to pass through its territory en route to Manitoba, forcing the regiment of roughly one thousand men to march and portage their way through the rough lands and lakes between the Lakehead and Fort Garry on the Red River. This major effort took place through an expanded Prince Arthur's Landing, which eventually became the City of Port Arthur. The rebellion was crushed, but the need for a modern transportation link between the Canadian Lakehead and the North West Territories remained unfulfilled. Passage between Manitoba and Ontario was largely conducted through United States territory, using the new railroads that were springing up all over the Midwest.

There was a growing demand in Canada for an east–west railroad link, particularly by British Columbia, which threatened secession if the link was not established. In the early 1880s, the Government of Canada, led by John A. MacDonald, authorized the establishment of the Canadian Pacific Railway, and work commenced on the massive and difficult project in three main areas: the mountain passes of the Selkirk and Rocky Mountains; the route across the Prairies to link what is now Calgary to the boom settlement of Winnipeg on the Red River, in what became the province of Manitoba; and a route through the seemingly endless morass of muskeg and lakes of the Canadian Shield, covered with evergreen forests stretching westward all the way from Sudbury to the shores of Lake Superior and then further west to Lake of the Woods and the Red River. The completion of this last route would be the making of the Canadian Lakehead.

However, the Canadian Pacific Railway almost went under several times in 1883 and 1884 as they ran out of money, and the Government of Canada dithered in indecision while contemplating the political effect of a further loan to the railroad. Then, in 1885, the figure of Louis Riel appeared on the scene again. Returned from the United States, he was back among his Métis followers now living in Northern Saskatchewan, close to the reservation lands of the Blackfoot and other restive tribes. Although the Métis and the Indians may have had some legitimate claims against the Canadian government, the path to outright rebellion was chosen by Riel and other leaders of the Métis and Indians. The Government of Canada reacted by sending a large military force from Ontario out to the West over the semi-completed Canadian Pacific Railway to join the outnumbered North West Mounted Police force in Northern Saskatchewan confronting the Métis, the Blackfoot, and the Cree. The completion of the east-west rail link was now the country's highest priority.

Small battles with the Métis took place at Duck Lake, Fish Lake, and Batoche, and with the Indians at Frog Lake, Fort Pitt, and Cut Knife. After some setbacks, the Crown was eventually victorious and Riel was executed. The parkland of the Prairies became ranch land and was put under the plough. The Canadian Lakehead began to grow as the city of Port Arthur expanded to meet the requirements of expansion in the west, and to serve the flow of trade in grain and other products that the railway brought to the shores of Lake Superior for onward transportation to the Lower Lakes and seaports along the St. Lawrence River.

In 1892, the city of Fort William was inaugurated on the banks of the river Kam, and the future of the Lakehead seemed boundless. Speculation in land was rampant, with all the resulting chicanery and skulduggery taking place that one could predict. A distant relative took part in the buying frenzy and, almost a century later, the record of his worthless purchase remains framed on the wall of his old home in England, arousing odd glances of interest from his grandchildren as they contemplate what might have been.

By the First World War, the waterfront of Port Arthur and the east bank of the river Kam were dominated by many large grain elevators and their accompanying turning basins and wharves. It was truly an age of expansion. The twenties brought further prosperity to both cities, enhanced by the illegal but lucrative liquor trade into the prohibition-era United States. Many a local fortune was founded in such a murky past. The thirties brought depression and great hardship for some as world trade diminished, affecting the movement of grain and lumber products through the Canadian Lakehead. Things did not improve until the threat of war in Europe sparked a revival that became a boom as World War II arrived and Canada was expected to play a major role in supplying fighting forces from a newly renovated industrial base, in addition to furnishing the food and raw materials to underpin the worldwide war effort. Port Arthur and Fort William benefited much from the tremendous increase in activity.

Now, in 1948, the pace had slackened somewhat, but jobs were still plentiful, particularly if one were prepared to work outside the cities. Accordingly, with a little help from family, I found myself employed as a warehouseman for the Abitibi Company in their base camp near a place called Dorion, southwest of Nipigon, serving the many operating work camps cutting lumber for the paper mills of the Lakehead and beyond. For a seafarer, it was a startling change of scene, and I had to learn a new vocabulary. Gathering supplies for distant camps was a challenge at times, as I wrestled with the names of pieces of equipment that were unique to some of the horse teams that still provided the bulk of transportation in the more remote camps around Lake Nipigon. My memories of childhood on family farms in Aberdeenshire did help somewhat. Food and clothing for the camps was an easy chore as long as you had a strong back. The nearest nightlife was in the village of Dorion, about fifteen miles away by bush road. A rather dingy hotel with an even more disreputable beer parlour provided the only attraction, although two women were reputed to ply another trade farther out of the village. One visit to the beer parlour was enough for me. A fight between the French and the Finns threatened to engulf me, so I decided that a dignified retreat was in order. This visit coincided with the first breath of winter—snow flurries, frost, ice—so I decided that I should perhaps reconsider my attempt to swallow the anchor and go find myself another job at sea.

I made my peace with my beloved in Port Arthur, and then headed by train for Montreal where I was sure that I would find a suitable berth. I checked in with Charlie and Bella Barber, my wife's uncle and aunt, who made me very welcome. The Canadian Merchant Navy Officers Guild arranged for my interview with the marine superintendent of Goulandris Brothers Ltd., and shortly thereafter I reported as second officer aboard the SS *Seaside* in the Montreal docks. She was a typical Canadian Park vessel of seven thousand gross registered tons. Her wartime name had been the *Sevenoaks Park*. Her Master was an ebullient Scotsman called Danny Scott. He had been in a German POW camp for several years after a German Raider had sunk his ship in the North Atlantic. That experience had certainly not inhibited him, and during the voyage to come he told us many stories of life in the camp, some of which I will divulge later in the story.

The first mate was an Englishman whose name I have forgotten, and the third mate was from Lotbinière on the St. Lawrence River. The radio operator was from the Renfrew area of Ontario and a good shipmate. His name, if I recall correctly, was Jim Schubert. I cannot remember the names of any of the engineers, but I do remember that the chief engineer had two very good-looking daughters who came down to watch us depart and wave goodbye to their dad. The chief steward was a smooth type from Montreal whose real character only came to the fore later in the voyage. The deck crew were mostly from Montreal, with a few from the Maritimes. They and their compatriots in the engine room and the steward department were all strong supporters of their union, the Canadian Seaman's Union (CSU). The CSU had an affiliation with the Communist Party and seemed to be constantly in strike action against the Great Lakes Shipowners. It was rumoured that such aggressive action was contemplated soon against the owners of the deep-sea fleet.

Seaside was bound mainly for the Indian subcontinent, and our cargo consisted of manufactured goods of various descriptions, but particularly steam locomotives, rails, and ancillary equipment. The locomotives were stowed on deck and were destined to be the source of some serious trouble on the voyage. We were to complete our loading in Portland, Maine. Our departure from Montreal was uneventful until the crew discovered that there was no ice cream on board! We then anchored in the river until this horrible mistake was rectified. The power of the CSU had been displayed. We then set off for Portland.

It was now past the middle of November and the signs of the coming winter were everywhere in evidence as we made our way northward into the Gulf of St. Lawrence before passing between Newfoundland and Nova Scotia, and then out into the broad Atlantic. The weather turned most foul when we headed southward into the Gulf of Maine, and before long we were encountering winds of hurricane force and huge waves that threatened to engulf us. It was among the worst storms I had ever been exposed to and it became quite terrifying as we found ourselves wallowing abeam the huge waves. *Seaside's* ten knots flat out were insufficient for a spell to allow us to keep her head into the sea and wind. At one point, I thought we were finished as I clung to the pedestal of the starboard wing gyro repeater with my body outstretched almost parallel with the bridge deck five feet below. It seemed to take forever for *Seaside* to come back from her roll in the trough of the waves, and this extreme movement was repeated a number of times before we managed to get out of the trough and put the wind and the sea on our port bow. By this time, our deck cargo was becoming unsecured and valiant deeds had to be undertaken by the crew on deck to lasso the most troublesome items and save the ship. A rolling steam locomotive of seventy tons can inflict enormous damage on hatch coamings and on bodies. Eventually, some semblance of order was achieved, and we were able to ride out the hurricane.

Needless to say, we were very happy to enter the harbour at Portland. For several days, we repaired the damage to the ship and deck cargo while loading additional general cargo into the holds. Most of the cargo seemed to consist of engines and engine parts, plus bales of clothing. However, there were many drums of volatile liquids that were stowed on deck amid the afore-mentioned locomotives. It was now almost December, and the local population was celebrating US Thanksgiving, although the stores were bright with Christmas lights and Santas paraded on every corner. I felt a bit homesick, particularly as the local radio stations conspired to feature the continual playing of a tune called "Slow Boat to China." That was us, even though we were bound mainly for India, and there was no way I could smuggle my nearest and dearest on board.

I think we found Portland rather dull as a port city. If there was a lively nightlife, then we failed to find it. Even a taxi ride to Bangor failed to unearth any delights that would interest the seafarer. So we sailed out into the Atlantic Ocean heading for the Mediterranean, conscious of our piety and good conduct, but fully intending to find and visit the fleshpots available in the Orient.

The weather at first was moderate and we charged along at full speed, but as we approached the area north of the Azores, the storm clouds gathered and soon we were battling another hurricane. We were able to avoid falling into the trough this time due to our having to deal with a more following sea, but the waves swept on board and undermined our deck cargo again, causing great confusion and movement on deck, particularly as many of the wire-and-chain fastenings parted under pressure, thereby allowing the locomotives and drums to move with each roll or pitch of the vessel. Two locomotives ran into the bridge housing and were secured there with all the chain we could find. The drums were lassoed as in a Wild West rodeo, and gradually everything came under control. Once again, a fine effort on the part of all concerned.

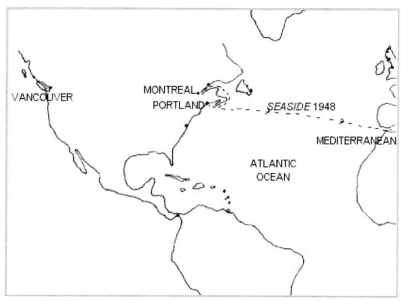

Figure 6 - Around the Globe voyage of *Seaside* Part I, 1948–49

The rest of the voyage across the Atlantic Ocean was uneventful, as was the passage through the Mediterranean to Port Said. During that period, I learned a bit more about our crew, particularly the able seamen and ordinary seamen and their petty officers, the bosun, and Chips, the ship's carpenter. They talked a great deal about their lives, their former postings, and their families, perhaps more so than the crew of a British-flagged ship, where the dividing line between officers and crew was much more rigid in structure. Some of these men had seen action in World War II, but most of them had been too young for service and had grown up in a post-war society that seemed to emphasize a split between those who held power and the workers—and they clearly identified with the workers. This attitude manifested itself in their strong allegiance to their union and its leaders' pro-communist stance in both pronouncement and in deed. I had nothing against a union, indeed I was a member of the Canadian Merchant Service Guild, which fought for better salaries and working conditions. However, the CSU was not just interested in the welfare of its members; it also had a political agenda that was being promoted vigorously on board the *Seaside*. Now, in all fairness, some of the older, more experienced men appeared slightly cynical about the politics and left it to the younger hotheads to spout their nonsense. The result could have led to a poorly disciplined crew in cases of emergency, but somehow the years of training became the glue that bound us all together, even in the worst of times. Nevertheless, the stories of breakdown of discipline in other Canadian vessels caused me to be thoughtful and watchful.

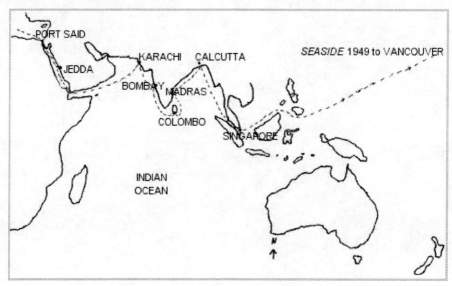

Figure 7 - Around the Globe voyage of *Seaside* Part II, 1948–49

We passed through the Suez Canal without difficulty, fending off the bumboats trying to come alongside and sell us women, carpets, baskets, local whisky, and Spanish fly. Additionally, slightly pornographic pictures were available, I assume well below the standard expected today by readers of Penthouse or viewers of any porno website on the Internet.

Upon exiting the Suez Canal at Port Tewfik, we headed down the Gulf of Suez toward the Red Sea and our first port of call, Jedda, in Saudi Arabia—the entry port for the Moslem faithful making their pilgrimage to Mecca. The approach to the port was difficult, with many partially submerged reefs guarding the anchorage off the city. There were small wharves alongside, but we were forbidden access. In fact, after the port authorities visited us, we were advised that no shore leave would be granted. A Saudi guard was placed on board, armed with an ancient blunderbuss—no English and we had no Arabic. He had to be fed on board, and somehow or other this good Moslem was persuaded, at least once, to eat veal which was really pork. The chief steward thought this was hilarious, but I think most of us felt rather uncomfortable and slightly ashamed at the subterfuge. The guard never knew that he had eaten pork and went ashore a happy man.

The cargo for Jedda was mostly automobiles and parts and, after a couple of days, we were on our way south down the Red Sea to the Straits of Bab el Mandep and the Gulf of Aden. It was now mid-January 1949, and I thought back to my time in cable ships around these waters with some nostalgia. Life on board *Seaside* was not all bad, but certainly did not measure up to the halcyon days of *Norseman*.

We passed fairly close to the island of Socotra, which lies east of the Horn of Africa in the Arabian Sea. It was a strange-looking place, all mud and terraced walls, presenting a face of menace to the passing seafarer. As the home of Arab pirates for centuries, it was still a place to avoid, even in 1949. From there, our route took us directly to the port of Karachi, which I had last visited in 1947 on board the cable ship *Norseman*. Then it had been part of British India, now it was the dominant city of the nation of Pakistan—newly formed in 1947 when the British left India. Many things had not changed—the smells, the huge population, the beggars, and the confusion of noise—but the port was obviously under the control of the Government of Pakistan, with locals running every aspect of commerce and politics. As Canadians, we were welcomed with open arms, even though some of us spoke with suspiciously British accents. What amazed me were the social functions to which some of us were invited, where many of the wives seemed to have abandoned the purdah and took part in the ongoing exchange of opinions and viewpoints as vigorously as their men folk. Truly, the world was changing.

Our cargo was discharged within three or four days, so we did not have much time for a run ashore, but the hotels and restaurants were well stocked with goods made in Europe and North America. Even Moosehead beer was available in some establishments. We departed Karachi, bound for Bombay, without further ado.

I have described Bombay in my book *Mandalay to Norseman*, particularly in 1947, shortly before independence for the Indian subcontinent. In 1949, not much had changed outwardly; cricket was played everywhere around the city, and military bands could be heard on the Colaba Causeway, but very few other signs of imperial power remained. A lot of British were still in India, but most of them were supporting the Government of India in re-establishing law and order in the aftermath of the horrendous events brought about by the partition. A visit to the Taj Mahal Hotel revealed few changes, many British military officers and their wives awaiting passage home, and a prominent sign over the entrance to the hotel barring South Africans from entry. This action was taken in retaliation for legislation taken against the Indian population of the province of Natal by the apartheid government of South Africa.

We discharged our cargo in Victoria Dock, where I had first visited in 1942 on board *Kindat*. The docks were as busy as ever, betel juice stains were everywhere, and a confusion of noise from voices speaking in the many tongues of India overlaid with the harsher noise of steam winches and other equipment filled the air. Amid it all, *Seaside* struggled to unload her cargo of locomotives bound for the railways of India. It was tricky work, as we deployed our own heavy lift derricks, but all locomotives were successfully placed upon the wharf, to everyone's relief.

Before leaving Bombay, I managed to visit my old haunt, the swimming pool at Beach Kandy. The weather was nice and cool and the water was grand, with lots of svelte young ladies to enliven the scene. A visit to the nightclub at Green's was also undertaken in the interests of nostalgia. They still required the wearing of neckties for entry into the establishment but, as usual, had a supply for rent for the evening for seafaring slobs like me. I found myself sitting with the lead singer of the orchestra, a lady from England, talking and drinking and passing the night away. I was obviously slipping, but the captain captured another young lady of obscure background and vanished into the night. Also present at our table that night was an officer of the Indian Special Branch who told us of the events in Bombay the night that Gandhi was assassinated in Delhi by a Hindu fanatic. Apparently, rumours immediately flew around Bombay that the British were responsible, and a huge mob of Gandhi supporters gathered around the Gateway to India and marched on the Taj Mahal Hotel, the symbol of British power, filled with many Britishers awaiting passage home. According to my Indian friend, only a strong reaction from the police and the military prevented a terrible massacre.

We departed Bombay for Colombo, Ceylon (now known as Sri Lanka), with a course taking us fairly close to the west coast of India. Indeed, we were so close to the coast that we soon ran afoul of fleets of fishing vessels and long strings of fishing nets running out from the shore into deeper water. We soon learned our lesson and set our course at a safer distance offshore. We entered Colombo Harbour in late January and soon learned that all was not well back in Canada. Thank goodness, this was not bad news about family, but information that the CSU was now on strike against the Canadian deep-sea fleet and that no one knew how this would affect the ships, such as our own, that were on foreign voyages.

Our crew seemed slightly perplexed by the news, but noises of militant action were heard coming from a small but vocal minority. They wanted to tie up *Seaside* immediately and even approached a Russian merchant ship tied up astern of us, seeking their support for such action. The Russians appeared astonished at receiving such an overture and hastened to advise our captain so. It did get a little tense on board, but Danny Scott kept his cool, and the situation stayed under control. We discharged our cargo relatively uneventfully, while being made aware of the ethnic split that had taken place in Ceylon, after it achieved independence from Britain, between the dominant Singhalese Buddhists and the Indian Tamils of the north. However, all was quiet on the waterfront. We sailed without further delay for the South India port of Madras, the very first foothold of British power on the Indian subcontinent.

Madras I recall as hot and sticky, and without an imposing harbour such as Bombay or Colombo. While discharging cargo, we eagerly awaited further word from Canada on the CSU strike and all its possible impacts on the lives of everyone on board. Our radio operator managed to receive short-wave broadcasts from the CBC in Montreal, and we were advised that the ship owners of the deep-sea fleet were trying to break the strike by signing agreements with the Seafarers International Union (SIU), a union until now based largely in the United States. Our crew were perturbed and anxious, but otherwise continued to carry out their usual duties.

Seaside's next port of call, and her last discharging port, was Calcutta. The port city was located at the head of the Bay of Bengal, but could only be reached by penetrating the river Hooghly, a delta arm of the mighty river Ganges. Calcutta is about one hundred and fifty kilometres up the Hooghly from the Bay of Bengal. In 1949, it was located largely on the eastern side of the river. The land is flat but wonderfully fertile. The city was large—four million population at that time—a spectacular mixture of extreme poverty and large pockets of wealth. It was a creation of the British East India Company, founded in 1696 as Fort Williams, but close to the fishing village of Kolikata from which its name was derived. It was the capital of British India until 1912, when the government of India was relocated to Delhi.

The city contained many fine examples of architecture, both Oriental and Victorian, with the Maidan surrounding the site of the original fort, providing a large open park area where the population could stroll alongside the river. The Victoria Memorial, made of marble and based on the Taj Mahal, dominated the skyline. The Shahid Minar, whose column reached forty-eight meters in height and provided a good view of the city, was located on the northern part of the

Maidan. The river pilots, who were by now all Indian, guided us to tie up to large buoys. It became a major chore to secure the vessel to the buoys by anchor chains fore and aft. The infamous tidal bore was anticipated within hours. When it arrived, *Seaside* floated safely, but the tidal bore, reaching seven feet in height, wrought considerable damage along the shore, where locals worked close to the river, promenaded on their way to somewhere else, or just carried out their daily ablutions.

The river was also notable for the number of bloated corpses, human and animal, that were swept past the harbour and down toward the sea. We did not remain at the buoys for very long because a berth was found for us in the Kiddepore Docks, enabling us to discharge the heavier items of our cargo. It was hotter in the docks and smellier, noisier, and dirtier—not a pleasant place to be. Outside the dock area, people lived in absolute poverty and despair, particularly the many beggars, the crippled, and the disfigured from disease. A most sobering sight!

There were a few places where a sailor could relax away from the ship: the Marine Club, run by the Flying Angel Anglican Mission, a quiet, cool place where the beer was very cold, the Grand Hotel, where the tariff was much higher but the ambience well worth it, or one of the several nightclubs, like Firpo's, where the prices were outrageous but the girls enticing. Otherwise, it was a hellhole with little to commend it. Additionally, the CSU affair back in Canada was rumpling our own disciplinary situation on board ship, which sometimes translated into confrontation and fisticuffs ashore. A black eye and sore bruises were my testimony to this spreading ill feeling among the younger, more violent members of our crew.

The port was thronged with other vessels of various nationalities, but dominating them all were the many ships of the British India Lines, which was still a power in the seagoing trade of the Indian subcontinent. I was always pleased to see these vessels, as they reminded me of wartime convoys and were a link to my mother's cousin, Nelson Jamieson. He had employed me as his game bag custodian when he went hunting for rabbit, hare, and pigeon over the extensive acreage of his mother's farm in Aberdeenshire back in the late thirties. He was now Commodore Captain of the British India Lines, and I eagerly sought him at the officers' club of the British India Line. Unfortunately, he was on a voyage and we were unable to re-establish contact.

The history of Calcutta and the surrounding province of Bengal I found fascinating. Before the arrival of the European traders, the area was under the control of the Mughal Empire for many years. The arrival of the Portuguese at the end of the fifteenth century was followed in the next century by the Dutch, French, Danish, and English. The English East India Company went on to become the most successful, spreading from western India into Bengal in the seventeenth century and establishing factories or trading posts in and around the river Hooghly. The relationship between the East India company and the Mughal emperor was excellent and much to the advantage of the company. But, as the Mughal Empire gradually weakened in the early eighteenth century and the Nawabs of Bengal became gradually more independent of the rulers in Delhi, a struggle for power developed with the East India Company for political control of Bengal. This confrontation came to a head in 1756, when the Nawab attacked and overcame the resistance at Fort William. The infamous Black Hole of Calcutta episode took place at this time, with many English prisoners dying while cooped up in a dungeon of the fort. The fort was retaken later that same year by Robert Clive and his troops from the Madras garrison. The battle of Plassey, fought later that year, cemented British control of Bengal, and made Clive a very wealthy man. This battle and the battle of Buxar in 1764 paved the way for almost two hundred years of British rule in India. Calcutta's population and the wealth of the city grew enormously as it became the seat of political power in India.

Seaside was unable to pick up cargo for the return passage while sailing around the Indian subcontinent. Apparently, our company was not a member of the proper trade association and was therefore denied access. What a crazy business, as the world was still trying to recover from the trade embargoes of WW II. So, we departed Calcutta for Singapore in ballast, with our eventual destination to be Vancouver, where we were expected to load lumber and wheat for the United Kingdom. The passage across the Bay of Bengal and down the Strait of Malacca was pleasant, with good weather prevailing. Our approach to Singapore was straight forward, but here I almost got the ship into serious trouble. The captain had left me no specific instructions regarding the final approach to the port, so reaching back in to my own experience in navigating around Singapore waters in cable ships, I took *Seaside* right into the approaches to the inner harbour, only to be told by signal lamp from onshore to back off a few miles and await the arrival of a proper pilot. I did so and then awakened the captain and confessed all. He was very understanding and took the full umbrage of the local pilot authority upon himself. A prince of a fellow was Danny Scott!

The pilot was English and had that superior attitude that can get right up a Scot's nose, which probably decided the captain in my favour. However, there was a notable difference in the efficiency of Singapore port compared with the sometimes close to chaotic conditions prevailing in the ports of the Indian subcontinent. Here, everything worked, even if it was in the pukka sahib manner. We were only in Singapore long enough to fuel and take on supplies, but even in that short time, some members of our crew managed to fall afoul of the law. After the easygoing dock police in Bombay and Calcutta, they found that the law in Singapore emphasized that, as white men, they were expected to set an example and, when they failed to measure up, the punishment was severe. It took us an extra day in port to pay their heavy fines and extract them from jail. They were a chastened group of individuals on the long voyage home to Canada.

While in Singapore, I visited the cable station that I knew so well from my days on board the cable ship *Recorder*, but there was no one around who remembered me—a slight blow to my high opinion of myself. After departing the harbour, we passed close to the Riau Archipelago, where I had first experienced the art of cable repair in relatively shallow water, and then headed out for the Horsborough Light and the South China Sea. We passed through the Balabac Strait between Palawan Island in the Philippines and British North Borneo (now known as Sabah in Eastern Malaysia) and on into the Sulu Sea. Our route then took us through the islands of the Sulu Archipelago, before crossing the Celebes Sea to the south of Mindanao. I recall being up on the monkey island while fixing the ship's position and realizing that the water was so clear I could see the outline of the sea bottom—a mixture of sand and smooth rock. I hurriedly rechecked the echo sounder for depth and was reassured to read in excess of ten fathoms, which checked out with the depths shown on the nautical chart. Whew!

Seaside then commenced her long journey across the Pacific Ocean to British Columbia. It was a very long voyage in ballast and against the seas and the prevailing winds. We bumped and ground our way day after day at an average speed of nine and a half knots. It was miserable, defeating even our efforts to spruce up the ship for arrival in Vancouver. Finally, in early April, we could see the mountains of Vancouver Island on the horizon, and made our way up Juan de Fuca Strait between the island and the coastline of the US State of Washington to the south. Everything looked so pristine and wholesome on shore, and we welcomed the pilot who boarded us off Race Rocks with some enthusiasm. We had been receiving news of events on the CSU strike front through our radio operator but, according to our newly boarded pilot, things were much worse than we had thought. The shipping companies were now using the SIU to force CSU crew members off Canadian deep-sea vessels docking in Canadian ports. In retaliation, the CSU was striking Canadian ships in foreign ports. It sounded like an awful mess.

Seaside proceeded through the Gulf Islands and the Strait of Georgia to Burrard Inlet, the entrance to the port of Vancouver. From there, we went no further, being ordered to anchor and await instructions. It was a strange homecoming, where we joined several other Canadian ships anchored in the vicinity of Point Grey and English Bay. Eventually, a launch took our pilot off and another launch arrived with the company agent on board. At last, the news filtered down that we would remain here for several days before going alongside to commence loading grain. Mail was distributed, and large quantities of fresh milk were handed out in cartons to the ship's company, which were much appreciated. Then a waiting game began, with a CSU launch passing messages to our crew by megaphone and an SIU launch slightly farther away broadcasting messages that were incomprehensible to those of us on board. The local newspapers—the *Province* and the *Sun*—were delivered on board and they were full of news about the strike. Indeed, other world news took a back seat to the action or perceived action on the Vancouver waterfront.

Finally, after several days of confusion, we picked up our anchors and made our way under the Lions Gate Bridge into the harbour, and then docked at the terminal wharves on the City of Vancouver side of the harbour. Those of our CSU crew who wished to do so left the ship in the late afternoon without incident. However, in the early evening the scene changed dramatically, with many police arriving on the wharf by vehicle and by launch, accompanied by a big contingent of press, armed with cameras and microphones. Something was about to happen—and it did! A large launch appeared alongside the wharf just ahead of *Seaside* and disgorged fifteen or so ruffians armed with pickaxe handles and other weapons, who rushed up the wharf to the *Seaside's* gangway and rapidly climbed on board. They were met by the two remaining CSU members still on the ship, who were all dressed up to go for a night on the town. Naturally, a fight ensued with the outcome never in doubt. Two beaten-up and bedraggled CSU members had to flee the ship, their home for the last five months.

I witnessed all of the foregoing and was thoroughly upset and confused. Obviously, the police and the press had been forewarned about what was going to happen, with the companies and the SIU well prepared to take action. No doubt it was considered necessary to take such drastic action, but it left a bad taste in my mouth. It was not my strike, but these men had been my shipmates for many months in foul weather and in fair, and I felt that they deserved better. To further emphasize the change that had taken place, two tugs immediately moved alongside *Seaside* and moved us from the wharf to the middle of Burrard Inlet, where we dropped anchor. We had been on shore power, so it took a while for us to get up steam and supply our own electric power.

I felt disillusioned by the events, and even after long chats with the captain and officials of my own union—the Guild—I was unable to come to terms with what I had witnessed. The SIU replacement crew were not bad fellows on the whole, but I was beginning to lose faith in the system. I felt that such action could only badly reflect upon the Canadian deep-sea fleet, and so it eventually transpired. Within a year, the fleet of one hundred and sixty vessels had been reduced to about fifty ships. I know that it was not only the CSU/SIU affair that ruined the fleet; lack of government support for fleet replacements, public ignorance of the important role of a Canadian merchant navy, and many other barriers, all played their part. But the Canadian government's role in assisting the SIU to oust the CSU was the crucial action that led to the demise of a deep-sea fleet.

We departed Vancouver in ballast again, but just bound for Victoria to enter the Esquimalt Graving Dock for essential repairs and to wait out the effects of the strike, which had by now spread to the stevedores in all the ports along the coast of British Columbia. I liked what I saw of Victoria, its beauty, and its strong British connection but, on top of the stress brought about by the strike, I was getting quite homesick and missed my wife a lot. Danny Scott, our skipper, had departed already, to take up a post in Montreal, so I decided to join him in heading eastward by train, but only as far as Port Arthur. Shortly after I left *Seaside*, the officer who replaced me was badly beaten up on board the ship in a dry dock confrontation between the two opposing sides. A sad ending to what had largely been a happy time.

Chapter Three

Sailing the Great Lakes and Other Canadian Waters

In 1949, the railways still played the major role in transporting not only goods but also individuals across the vast dominion. So I traveled by CPR steamer from downtown Victoria to downtown Vancouver, and then embarked on the CPR train that was scheduled to leave Vancouver every evening bound for Ontario and Toronto. It was a long train with many coaches and sleeping compartments. I was located in one of the standard berth sleeping cars, comfortable enough at night and with a good view of the passing countryside by day. Leaving Vancouver at night, we missed some of the spectacular scenery associated with the Fraser Canyon and River, but did see much of the grand views around Rogers Pass and the mountains surrounding Banff. It was a worthwhile journey, even as the train left the Selkirk Range and rumbled out into the high plains leading to the city of Calgary. I chatted with a few fellow passengers, particularly in the dining car, which I thought was a highlight of the trip, serving excellent food with good and courteous service. However, I cannot recall anyone in particular, so I suppose I did not make much of an impression on my fellow travelers either.

The rail journey beyond Calgary was across the Prairies, past places with names like Medicine Hat, Moose Jaw, and Swift Current, all guaranteed to stir my imagination and desire to know more about Canada. It was now late April and the ground was still covered in patches with snow and ice, but in the valleys one could see the first signs of spring approaching. We must have passed through Regina, but the next city I recall was Brandon, Manitoba, and before long we arrived in Winnipeg. Here we changed engine equipment and spent some time being shunted around before we were once more on our way. There had been little time to see anything of Winnipeg, then the largest city by population on the Prairies. A cool wind did not encourage much exploration. We were now approaching Kenora and northwest Ontario, where the land changed dramatically from the flat, often-treeless prairie to the Precambrian rock of the Canadian

Shield, with many small lakes and rivers, overlain with evergreen forest. It had been forty-eight hours since we had left the Pacific coast and I, for one, was getting tired of train travel and longing to see Port Arthur.

After a short stop in Fort William, we proceeded on to the station at Port Arthur, to be met by wife and other relatives and carted off home for a good rest. Doreen was working for Marshall Wells, a local hardware company, so it behooved me to seek employment as soon as possible. However, there was a chance to see the junior ice hockey playoffs, which was my introduction to the madness and excitement that attends Canada's national sport. To have the home team win by one goal in the dying seconds of a game was truly a great thrill to me and, I am sure, to all those present. To have one female supporter of uncertain age divest herself of her underwear and throw it on to the ice was an eye opener to the passion that ice hockey commanded of its more devoted supporters. I believe that was the year the Port Arthur Bruins won the Memorial Cup, emblematic of Canadian Junior Hockey supremacy.

A quick look around the Lakehead confirmed that work was getting harder to find ashore, although there were still jobs to be found out of town in dangerous occupations such as high rigging. Then I heard of a vacancy on a canaller that had just arrived in Fort William on her first voyage of the season. She needed a second mate so, on May 6, I signed on the SS *Shelton Weed* and sailed for the lower Lakes with a full load of grain. She was much smaller than the upper lakers, to enable her to traverse not only the Sault and Welland Locks, but also the twenty-seven tighter locks stretching from east of Kingston to Montreal. These locks bypassed the St. Lawrence River rapids and kept the port of Montreal and other downriver ports supplied with grain from the Lakehead, bound for the ports of Western Europe.

The *Shelton Weed* was owned by the Upper Lakes and St. Lawrence Shipping Co., based in Toronto. She was a vessel of around 1750 tons gross and had a total crew of twenty. The ship kept three watches, but all hands turned out for warping through locks and docking and undocking. Accommodation was tight but livable, while the food served was excellent and well prepared by the cook and his assistant. There was no uniform worn on board. Everyone, including the captain, seemed to be wearing out their old clothes.

I soon discovered that my deep-sea background was viewed with suspicion and that my navigational skills were seldom allowed to flourish. My duties were confined to watch keeping on my own in fairly open waters but shadowed by the captain in the closer waters of the St. Marys River or the Rivers St. Clair and

Detroit. In the locks, I worked alongside the other seamen, handling the mooring winches and the wire hawsers as required. We worked our way through the Welland Canal and the many locks stretching from Lake Ontario to Montreal, where we unloaded our cargo of grain.

Figure 8 - SS *Shelton Weed*, 1949

Then came a surprise. We were ordered to proceed further north on the St. Lawrence River to the vicinity of Tadoussac at the mouth of the Saguenay River, where we would prepare to load a cargo of pit props for Ontario ports. It was the month of May, and much fog lay over the St. Lawrence valley, particularly in the confluence of the two rivers. In addition, we were not fitted with radar and lacked a gyrocompass. In the circumstances, we spent a great deal of time at anchor while anxiously listening for the sounds of approaching traffic. At long last, we were able to proceed alongside to load our cargo, and then make our slow and cautious way southward toward Grosse Isle and Quebec City. The fog then cleared and we sped through Lac St. Pierre on our way to the Lachine Canal and onward through the other locks to Kingston on Lake Ontario, where we discharged our cargo.

It was then a long haul back up through the upper Lakes to the Canadian Lakehead for another cargo of grain, this time destined for the port of Three Rivers on the St. Lawrence River. We were only alongside at Fort William for a few hours, so I did not see very much of my wife. The life was hard and the days were long, but on the whole I was enjoying myself. The weather was still cool on Lake Superior and in the St. Lawrence River, but it was starting to warm up in the lower Lakes and promised to get much hotter. The fellow members of the crew did not act like the professional seamen I knew from deep sea, but they were efficient in all the tasks that they were required to carry out. Orders and casual conversation were delivered and received on a first name basis, except for the captain who was addressed as "Cap." Like most of the others on board, he lived in the St. Catharines, Niagara Peninsula, area of southern Ontario. However, we had crew members depart without warning while locking through, and we began to pick up replacements in Quebec who had little understanding of the English language. Meanwhile, the captain controlled our fates from the wheelhouse, where he sat on a large stool and pontificated on every subject under the sun. He was a big man, well fed and well proportioned, and he seemed to fill the wheelhouse to capacity. He was not well informed about international affairs but knew the local political scene very well. I looked forward to chatting with him, as I had not managed to find anyone else on board that enjoyed a bit of verbal sparring. I casually explored with him the prospect of my transferring to one of the company's upper lakers, where I might get the occasional opportunity of visiting the Lakehead. Things then happened with astonishing rapidity.

It was mid-June, and *Shelton Weed* was traversing the Welland Canal in the vicinity of Port Colborne. Cap advised me that he would forward my request to company headquarters in Toronto, but pointed to an elderly man walking along the wharf and suggested that I talk to him about my request. The elderly gentleman turned out to be Scott Misener, the co-owner of Colonial Steamship Co. Ltd. based in Port Colborne. He told me to go to his office for an interview that morning, which ultimately led to my departure from the *Shelton Weed* and my arrival in Buffalo, New York, to join the *SS Royalton*, a large upper laker discharging iron ore from Superior, Wisconsin. A new, demanding experience was about to begin.

Royalton was tied up alongside the steel works at Tonawanda in the north end of Buffalo on the Black Rock Canal. She was half empty, but her cargo of iron ore was being rapidly unloaded with grabs operating in every compartment. She displaced in excess of seven thousand gross tons and was capable of a good twelve-knot speed fully loaded. Her crew complement was thirty souls, but I soon discovered that most of the time she ran undermanned. As one example, there should have been three mates employed, but the third one never appeared. I found myself on six hours on and six hours off watch keeping, supplemented by several hours extra each day when warping through locks, loading or discharging cargo, or conducting clean-up operations in the holds preparatory to loading another cargo. But more on all of that later!

The captain was a different man altogether from the easygoing skipper of *Shelton Weed*. His name was Jim Walton and he was Commodore of the Colonial Steamship fleet. He was always in uniform, stiff necked, pugnacious, and articulate. He had personally battled the union on numerous occasions and had won every battle—no doubt with the aid of company-paid goons. Nevertheless, he was a formidable fellow. He so dominated his ship that now, after these many years have passed, I can scarcely recall the names of other members of the crew.

Figure 9 - SS Royalton, 1949

The accommodation on board was quite spacious and comfortable. There were even a couple of large suites where influential passengers could be entertained. The food was excellent, with fresh fruit and vegetables and dairy products always available. It was, however, a hard-work ship, and liquor and other temptations were banned. There were several women on board who were employed as cook, baker, and stewardess—a bright factor in the grinding day-to-day activity that was *Royalton's* reason for existence. We departed Buffalo bound up Lake Erie for Lorain, Ohio, to load coal. The loading operation was an eye opener to me. The cargo awaited us in chutes, each containing a railway carload of coal. We manoeuvred alongside, placing our open hatches under the chutes. One man in a shiny new Buick drove up, dressed in a spotless white "boiler suit." He then proceeded electrically to control the entire operation of loading our ship in concert with the mate on watch, ensuring that stability of the vessel was maintained. In less than three hours we had departed for sea, and the man in his Buick had driven home. Having loaded coal in many ports of the world—in Asia, South Wales, South Africa, and even in Norfolk, Virginia, amid clouds of coal dust, screaming foremen, and tired and filthy workers—I was most impressed by this almost silent demonstration of efficiency.

We were now bound north up the Detroit River past the cities of Detroit in the US and Windsor on the Canadian shore. In those days, the Canadian shore could easily be followed by the many Union Jacks flying in defiance of the Stars and Stripes on the opposite shore. Our route took us through Lake St. Clair and into the St. Clair River. We then set course up Lake Huron bound for Blind River in the Algoma District of the province of Ontario. Blind River was a coaling station for the Canadian Pacific Railway and was surrounded by a native reservation. It was located in the North Channel above Manitoulin Island. Discharging our cargo was a more laborious business than the high-tech operation seen in Lorain, but twenty-four hours later we had completed discharging and were on our way to the American Locks at Sault Ste. Marie. However, the trip to Blind River was relaxing for our overworked crew, and booze and pliant females were part of the attraction.

Our next port of call was to be Port Arthur to load grain for Port Colborne. We had a good run up the St. Marys River to the Soo Locks, where personal and ship's laundry was apparently exchanged on each voyage, together with fresh supplies for the galley. The speed with which we entered and departed the locks while conducting as much business as possible was quite remarkable. In a very short space of time we were heading up into Lake Superior and its cooler waters, en route for Port Arthur. I was able to alert Doreen regarding my arrival, and because I was in port for just a few hours, she came down to meet me. Naturally, I dragged her off to my cabin for a spot of lovemaking, to the crew's amusement and Doreen's embarrassment. It was too bad that they were unable to delay loading for a few more hours. It was now early June and I did not realize that I would not see my beloved again until late November. We sailed shortly afterwards with a full cargo of grain. The weather was perfect but got quickly warmer as we approached the St. Clair River. Once again I found my navigational skills and ship handling capabilities ignored, and was truly on my own on the bridge only when we were out on the open lakes. Frustrating, to say the least.

We discharged our grain cargo at the Port Colborne grain terminal and headed back up Lake Erie, awaiting further orders. Much to our surprise, we were to tie up in Sarnia on the St. Clair River and pay off articles. I packed my shoregoing gear and headed for the railway station, along with most of the crew, only a few remaining on board as a skeleton crew. Then, just as we were lining up to buy our train tickets, word came to turn around and rejoin *Royalton*.

Cargoes had been identified and we had a full season to look forward to while plying the Great Lakes. Our next port of call was Lorain, where we again loaded coal for the CPR depot at Blind River. From Blind River, we headed for the US

Lakehead in Lake Superior. Our destination was Superior, Wisconsin where we took on a full cargo of iron ore for Cleveland, Ohio, on Lake Erie. Cleveland in those days was a very polluted port, with much scum and debris defiling the waters of the harbour. Indeed, it was so bad that it had burst into flame on more than one occasion in the previous two years. We got to our berth without problem and quickly discharged our cargo before heading along the coast to Lorain for another load of coal for Blind River.

This routine continued throughout the summer and autumn, except that Buffalo became our main destination for cargoes of iron ore from Superior. It was a hard slog, with little time for relaxation. I found the long hours of work every day very tiring, and with little to look forward to. One small break in the routine came while in Buffalo, when an action of the US Government forced us to take a few hours off work. It was the enforcement of the McCarran Act, the registering of aliens (us) as a means of underlining the struggle against the perceived Communist menace. What a performance it turned out to be. We were bussed from Tonawanda to downtown Buffalo, where we were left to our own devices to find a suitable photographer to snap our scowling faces and attach them to an official document that announced to one and all that we were "Alien." I felt as if we had been labelled "Martian" or some other suspect breed, as also did the photographer, who was obviously of mid-European background and, by his expression and mutterings, quickly conveyed his direst suspicions of our bona fides. I had become quite fond of Americans, but this overreaction to a perceived problem tempered my enthusiasm for all things south of the border.

It was now early October and I had already had a few run-ins with the Old Man about the lack of a third mate and the complete lack of adequate compensation to partially remedy the matter. An equipment failure at the unloading berth in Tonawanda gave us a welcome break from routine, and I took the opportunity, along with others, to head for Niagara Falls, New York, for some well-deserved relaxation. I had a wonderful time, my first drinking session in months, and I am afraid that I got rather snookered! I therefore had some difficulty in navigating the ship's ladder while under the stern gaze of Jim Walton. Nothing was said at the time, but the following day, when underway on Lake Erie, he proceeded to tear a proverbial strip off me. Most of it was well deserved, but he went a bit too far and I counterattacked, and so the die was cast.

We were bound for Port Arthur to load grain, so I prepared to sign off articles upon arrival. It was good to be going home, but I was sad at leaving *Royalton* because I knew that I had to now make another effort at swallowing the anchor.

My temper and my occasional binge drinking were worrying me. I needed to face up to my weaknesses and deal with them if possible.

Captain Walton and I parted company with a truce declared between us. It was to be almost one year before I would see him again, when he recommended me for a berth on another vessel of the Colonial Steamship fleet. In the meantime, it was to be that wrestle with the anchor.

Chapter Four

Swallowing That Blasted Anchor Again

My plans were to find a job as soon as possible—literally anything to keep me going, yet stay at home with my wife. I soon found myself employed at the Abitibi paper mill in Port Arthur, sorting logs in the mill pond on their way into the mill for processing into paper products. It was a job requiring mostly strong arms and legs and little in the way of brain power. I enjoyed the job until the weather turned colder, and then found myself working for an Ontario Hydro crew that kept the hydro line's right-of-way clear of undergrowth. We worked the line from Nipigon to Port Arthur in increasingly colder weather, accompanied by the usual snow storms and ice. The pay was pathetic, but transport to and from work was provided, together with a good hot meal at the nearest countryside café or restaurant. The crew were a good bunch of fellows, not too knowledgeable about the outside world, but with lots of common sense about matters affecting them at the Lakehead. One chap I remember particularly was Duncan McKay, who had a broad Scottish accent. It turned out that he was born and bred in Port Arthur and had never been any closer to Scotland than his upbringing by a Scottish mother and father.

My wife and I were staying with her parents in Port Arthur. They had purchased a small house on the outskirts of the city, and we lived in the attic. They were very good and understanding, but the circumstances gave us little chance to get to know one another again. At Christmas, the hydro crew was laid off and we now had to rely on my wife's salary. I sought other work, but there was nothing available, as demand for wood dried up and the shipyard sought new contracts. As we entered 1950, I began to wonder if our move to Canada had been wise.

With time on my hands, I thought a great deal about my life and my goals. There were breaks in the routine, such as visits to the Legion to imbibe large quantities of beer in convivial company. I had become a member of the Legion at the instigation of my brother-in-law, Bobbie Guerard, who had been a wireless operator air gunner with the Royal Air Force's Coastal Command. His aircraft had been shot down off Portugal; he was rescued by Portuguese fishermen, and then interned in Portugal while recovering from his wounds. He introduced me to his pals at the Legion, a rather wild but odd bunch. It was never dull in the Port Arthur Legion. There were still a number of Imperials in the Legion in those days—Imperial being the designation given to those who had served in the British Forces in the Great War. Their accents made me feel right at home.

My introspection caused me to take a look around me in search of an anchor. I was jobless, I was a spasmodic drinker, and I seemed to lack purpose. My wife, on the other hand, was obviously sustained in her life's purpose by her devotion the Catholic Church. Perhaps I might draw on this strength. So I set about to find out if the Catholic Church could be my salvation.

I already attended church with my wife on a regular basis, but I was only going through the motions and hoping thus to please her. As a Protestant, I had had to agree to take instruction in the Catholic religion and to have our children brought up in the religion if I wished to marry her. I had agreed, and my instruction was provided by a Father McClinchey back in Rock Ferry, England. He was from the north of Ireland, a tall, spare, bent man with the map of Ireland visible on his features. I found him very pleasant, if a bit strange. I was a skeptic and he tried to make a believer of me by citing what he maintained was a true experience. He was cycling home at night in Ireland when he was confronted by the Devil cycling the other way. He vividly described the light shining from the Devil's lamp and his own lamp lighting up the scene. I was thunderstruck that this priest could believe such nonsense and left him with many questions in my mind. When I had told my future mother-in-law of what had occurred, she laughed outright and noted that the Irish were a superstitious lot and I was greatly reassured. Now, I intended to find out more about the faith that apparently sustained so many.

The center of Catholic faith in Port Arthur was St. Andrew's Church, an imposing building from the outside and with an equally attractive interior. Neither ancient nor modern in its décor, its Stations of the Cross were eye catching, and its entranceway was dominated by a fine painting depicting the Jesuit Martyrs of Penetanguishene. It was a Jesuit parish with five or so priests of varied attitudes and views, which they mostly kept from their parishioners. There was a Scot from the Outer Hebrides who always spoke in whispers, a great bully of an Irishman who could be heard long before he sailed into sight, an Englishman with a supercilious sneer, and one of Italian background and full of piety. However, the most interesting priest was Father Bathurst, who was the pastor and administrator of St. Andrew's Parish.

When I made myself and my needs known to Father Bathurst, he welcomed me with open arms, perhaps seeing in me a dyed-in-the-wool but troubled Presbyterian, and therefore worthy of his attention and the display of his Jesuitical skills. Then again, perhaps I am wrong and he welcomed me from the fullness of his heart. No matter, he was interested enough to hear me speak of my concerns, personal and marital, and to quietly interject his own thoughtful comment from time to time. I met with him several times each week over a period of roughly two months, and began to seriously consider conversion to the Catholic faith. However, I was still troubled by the difficulty that I had in coming to a full belief in the faith. My rational mind could not accept the transformation of the bread and wine into the body and blood of Jesus Christ. Nor did I find it easy to accept the idea of papal infallibility in all matters of religion. Obedience had never been my strong suit, and now it was being tested as never before. The Church's official views on matters of social and political concern also caused me some uneasiness. It was the time of the McCarthy witch-hunts in the USA, and I discovered to my horror that the clergy appeared to support such excesses. I was not naive about the real threat of worldwide Communism, but questioned McCarthy's methods. All of these things combined to underline my confusion. I did go through the motions of becoming a Catholic, to please my wife and Fr. Bathurst, but they and I knew that my heart was not in it, and that I would have to find my own way. I would continue to attend church and would support my wife in bringing up our children, when they arrived, and that would be it.

Meanwhile, two months of 1950 had passed and I was still without a job in an area where the labour market had dried up. I then made another big mistake. I allowed myself to be persuaded that I could sell health and accident insurance going from door to door. What a disaster! The company was the Mutual Benefit Health and Accident Insurance Company of Toronto, a subsidiary of Mutual of Omaha, whose advertising messages could be heard from time to time on American radio stations. I was interviewed by the local manager, a chap called Johnson from Fort William, and his superior up on a business visit from Toronto. They seemed to feel that I had the makings of a good salesman—ho ho ho! They did, however, pay my way to visit Toronto for indoctrination, and I brought Doreen along for the change of scenery.

Toronto was still very much a provincial capital in those days. Although not dominating on the national scene, it certainly was the focal point of the economy of the province of Ontario. We were hicks from the sticks and made to feel so. The Mutual Benefit office was in an imposing building on University Avenue—a grand thoroughfare that led to the Ontario Legislature. Elsewhere in the downtown core, the signs of the massive construction of the Toronto Transit Commission's underground railway system were everywhere. A new era was a hand. Heavens, the first cocktail lounge in the city had opened its doors—Toronto the Good would never be the same!

We stayed at the Prince Edward Hotel, which was elegant but a bit shabby then. In addition to the indoctrination, we were taken out for dinner and then went to the theatre to see Tallulah Bankhead in the Noel Coward play "Blythe Spirit." It was worthwhile seeing the performance, although Tallulah was getting rather long in the tooth. All in all, it was a nice break from Port Arthur, which was still in the grip of deep winter.

After I returned, it was impressed upon me by the local manager that I had to get out and sell insurance as soon as possible to pay for my training visit to the big city. I started a correspondence course that was apparently designed to make me a better and more motivated salesman. Thus armed, I set forth to make my first cold calls on the unsuspecting population. I did not have a car, so I trudged Port Arthur's slushy/icy streets, utilizing the local bus service to concentrate my efforts in the more affluent-looking areas. It was tough—most people did not deign to answer the doorbell, or when they did would keep me at bay on the doorstep and then dismiss me from the premises. Very occasionally, I would be invited indoors out of the cold, and given a real chance to make my pitch.

I soon recognized that people were dubious about the product I was selling. There were many exceptions in the fine print of the policies that could negate the value of the product. Many people really needed the coverage, but their health made them ineligible. Some of my most interesting discussions were with older people in poor health. Their lives often had been very full and happy, but their future did not look very bright. In doubtful cases, I could not lie just to sell a policy and so left without that sale that other salesmen would not have hesitated to complete. I was not a good salesman, but perhaps I could have survived if I had believed in my product. Some days, I had to take time off to get away from the soul-destroying routine, and sought shelter in a movie theatre where the film and the warmth would provide temporary solace.

There were occasional highlights when making my calls. I once found myself in an illegal drinking establishment and managed to sell a substantial policy to the owner, who had formerly been the Madame of a "house of ill repute" in Winnipeg. I considered that day to be a triumph! Another was a visit to the home of an ardent female evangelical, who could not make up her mind on whether to buy a policy from me, convert me, or seduce me. I will leave you, gentle reader, to draw your own conclusions.

As the weather brightened into spring, I teamed up with another salesman who had a car. We were thus able to visit potential clients in the hinterland around Thunder Bay. The clients were mostly farmers and wood lot operators who were interested in purchasing accident insurance. The other salesman was a chap called Joe Doncaster, who was a native of Fort William. He was a good companion, but very cynical about the health and accident insurance business. He had been a dispatch rider in the Canadian Army Signal Corps in Northwestern Europe in 1944 and 1945. His tales of the Allied thrust for Nijmegen and Arnhem were fascinating to me. His wife was an English war bride, and his sister was married to one of the Chapple Stores family in Fort William. We fairly rapidly covered the area looking for potential customers, and turned our attention again to the twin cities.

Then, on the 25th June 1950, the great big outside world intervened—and our little world at the Lakehead was almost immediately affected—by a military action taking place half a globe away. The North Koreans had suddenly invaded South Korea, and the United States, and soon the United Nations, were committed to repelling the invaders. Canada immediately began to re-arm and pledged forces to fight in the Korean War.

The seeds of the Korean conflict were sown in the aftermath of the Japanese surrender in 1945, when the United States and the Soviet Union agreed to a purely military demarcation of Korea on the 38th parallel to facilitate the surrender of enemy forces. Korea was supposed to eventually become free and independent. By September 1947, talks between the US and the Soviets had reached an impasse on the subject of Korea, and the US turned the problem over to the United Nations. The issue of Korean independence became another irritant on the world stage between the two powers. Most members of the UN favoured all-Korean elections, but the Russians refused, claiming that only the powers that had accepted the Japanese surrender should decide on the fate of Korea. In 1948, separate republics were declared in both South and North Korea. Now, two militarily antagonistic states faced one another along the 38th parallel. Sporadic fighting was soon reported but, even so, the US withdrew its forces from the South in 1949, in the belief that Korea was of only peripheral importance in the event of a global struggle. It therefore came as a complete surprise when North Korea invaded the South on the 25th June 1950. The invading army of seven infantry divisions and one armoured division, together with amphibious landings and air attacks, soon captured Seoul and swept on to Taejon.

In early July, US troops from Japan returned to Korea, and the United Nations were asked for help to repel the invader. The UN Security Council, in the absence of Russia, called for the complete withdrawal of North Korean forces on the 27th July, and several member states of the UN immediately sent troops to South Korea to aid the Republic of South Korea and the US forces, who were by now hemmed in at a narrow bridgehead at Pusan on the southern tip of the Korean Peninsula. Canada prepared to send an infantry brigade and armoured group to Korea to join the Commonwealth Division, which was already in action at Pusan.

At the Lakehead, there was no immediate effect militarily, but business began to pick up steam and jobs became more plentiful. Plans were afoot to start building minesweepers at the shipyard, and new assembly lines were contemplated for the construction of military training aircraft at the Canadian Car & Foundry Company's plant in Fort William. In the new atmosphere of expansion, the painful pursuit of health and accident sales began to look increasingly futile. So Joe and I terminated our brief partnership and began to look for other occupations.

The very first work that I found was unloading box cars filled with grain at the United Grain Growers' elevator in Port Arthur. It was, I believe, the hardest work I had ever undertaken. Muscles screaming, choking clouds of grain dust, it was a nightmare! Most of the other members of my team were far fitter than I was—the salesman days had taken their toll—and I found it hard to keep up. I quit after one week and was immediately hired by the shipyard as a fitter's helper, which was much more to my liking and required little effort except to remember to always look busy and occupied with the job at hand. There were several minesweepers under construction, built with wooden hulls and aluminum framing, an answer to the magnetic mine menace. I trailed my fitter around the vessels and the workshops while he made measurements for various assemblies, and assisted him as required by supplying tools as necessary and arranging for welders to join his metal creations together. He seldom stopped talking, and I was therefore the unwilling recipient of much of his family history and its present strains and strifes. However, he was a cheerful little man, and patient with me and my inexperience.

The shipyard experience came to an abrupt end when I was offered a job as mate on the upper laker *Viscount Bennett*, operated by the Colonial Steamship Co. She was an elderly vessel of six thousand two hundred gross tons propelled by the usual reciprocating steam engines, and the job would only last until the mid-December freeze up, but I was becoming anxious to regain my sea legs or, as it was, my lake legs. The captain was a nice older fellow called Charlie Cole who treated me well, even letting me do the odd bit of navigation on my own. We sailed from Port Arthur with a full cargo of barley bound for the breweries at Milwaukee on Lake Michigan.

The weather was foul on the voyage down Lake Superior to the locks at Sault Ste. Marie. It was now late October, and the Lake was living up to its reputation as a devourer of ships and those who sail in them. We survived the passage and entered the locks without too much trouble. The usual quiet display of efficiency prevailed—food supplies, laundry, clothing replacements and ship's garbage—all arranged and disposed of before the lock gates opened to send us on our way down the St. Marys River. Impressive to one brought up in the shouting confusion of similar operations in the UK and the Orient.

There was a US Coast Guard station farther down the St. Marys River that wanted to know who we were. Instead of pronouncing the word "Viscount" in the British or Canadian manner, the radio operator insisted on identifying our vessel as "Vissscount Bennett," to much hilarity on board our ship. We were allowed to proceed without incident after this international exchange. The weather was brisk but clear as we passed through the Straits of Mackinac and entered Lake Michigan, but deteriorated into sleet and driving snow as we headed for the port of Milwaukee, Wisconsin.

Milwaukee harbour was dominated by large grain elevators, and one of those was our destination. We discharged our cargo at a fairly leisurely pace, thereby giving us a chance to see the city and partake of its attractions.

Actually, from what I remember, it was a typical port city, with lots of bars close to the docks. Somewhat European in atmosphere, befitting the city's population of large numbers of Poles, Germans, and other middle Europeans, the bars were a mix of the lively and the more sedate, serving European-style food and, of course, copious amounts of beer. They were friendly places where you could converse with the locals or indulge in a game at the billiard table or the shuffleboard. Pleasant places to relax when off duty. There was, however, an underlying sense of concern as one listened to the radio reports of events in faraway Korea and absorbed the comments and expressions of dismay conveyed by the locals.

All was not going well in Korea. After defending the Pusan pocket and conducting a massive sea landing at Inchon that outflanked and destroyed the North Korean army, the UN forces under US command swept north across the 38th parallel, occupying Pyongyang before pressing farther north to the border with China. In late October 1950, large formations of Chinese troops were detected in North Korea and, in November, the Allied push to the Yalu River and north of the Chosin Reservoir ran into fierce Chinese resistance. It soon became apparent that the Allies were overextended and in great danger of being enveloped and defeated. This appalling news was just beginning to filter into the minds of the regulars in the Milwaukee dockside bars we were frequenting. Much worse was, of course, to come.

We completed discharging our cargo of barley and headed up Lake Michigan to the Straits of Mackinac and the entrance to the St. Marys River. We were in ballast and bound for Fort William for another cargo of barley. It was starting to get cooler at the Lakehead, with light ice forming on the River Kam, but we proceeded to load without difficulty and were soon on our way back down Lake Superior to the Soo Locks. Once more, we set course for Lake Michigan, but this time our destination was the small port of Manitowoc, located about eighty miles north of Milwaukee on the Wisconsin shore. It was a quiet place, with none of the attractiveness of the Milwaukee harbour. We discharged our cargo in two days without incident and headed once more in ballast back up to the Lakehead.

I must comment upon the great advantage that enabled these upper lakers to navigate even in the thickest fog on the lakes. Together with gyrocompasses, the vessels were equipped with radar to add one or two or more cargo-carrying voyages per season. A great boon to their owners! The canaller I had sailed on previously had not been fitted out with radar or gyro and was delayed by thick fog for several days in the St. Lawrence River. Indeed, none of the ships I had sailed on before coming to Canada had been so equipped, not even the cable ships, where accurate navigation was absolutely essential.

The weather was fouler than ever on Lake Superior, and the ice was thickening in the Kam River when we loaded our next cargo. We were bound for Buffalo, New York, with a cargo of wheat. Once we cleared Lake Superior, the weather moderated, and we had a fast passage down Lake Huron, past Detroit, and then along Lake Erie to Buffalo. We docked at our elevator around midnight but, thank goodness, the work crew was not prepared to commence discharging our cargo until a more civilized hour. However, there was a bar just a few yards from the elevator which, I was astonished to find out, stayed open until 3:00 A.M. A couple of beers went down smoothly. In addition to the barkeeper and me, there was only one other customer—a salesman type who entertained us by telling tall tales of his experiences and discussing alcoholic drinks and their potency. He insisted that the one drink that always gave value for money was Scotch—stating emphatically that there was no such thing as a bad Scotch. After receiving such wisdom, I went back on board *Viscount Bennett* for a well-earned rest.

Discharging cargo in Buffalo was uneventful, and we headed once more for the Canadian Lakehead in ballast. All the cities and towns along the shores of Lake Erie and in the Detroit and St. Clair Rivers were bursting out with displays of bright lights—Christmas was surely coming. The food on board, always good, became quite outstanding. A French-Canadian couple did a superlative job of keeping us well fed. We loaded wheat in the Kam River, where I was able to see my wife briefly before heading out into Lake Superior. This time, we were bound for the elevator at Kingston, Ontario—a long run down the lakes to the Welland Canal and then across Lake Ontario to the Thousand Islands where the St. Lawrence River proper begins. The discharging operation was by day and it was a relatively slow affair, leaving us tied up alongside for a couple of nights. The opportunity to see Kingston, one of the older cities in Canada, was available, but I joined my crew mates instead to visit a small watering hole across the river in New York State. It was now early December, and lay-up time for *Viscount Bennett* was fast approaching. Upon completion of discharging, we returned to the Lakehead for our last cargo of the season. The weather was now getting much colder, with light ice forming in the Welland Canal locks and at Sault Ste. Marie. Snow flurries gave way from time to time to heavy snow falls—after all, it was now well into December. I was expected to remain with the ship upon completion of loading in Port Arthur, and return with her to Port Colborne on Lake Erie, where *Viscount Bennett* would lay up during the winter months. However, I was given the option of signing off the ship in Port Arthur, which would save me the cost of a rail fare home from Port Colborne. I gratefully took the option after saying farewell to several people I had really enjoyed sailing with.

I was welcomed home by my wife, who was already making preparations for Christmas. I also found myself working in the local liquor store, helping to keep the local population well supplied with their tipple of preference. It was perhaps a fitting end for what had been a year of ups and downs, with me still trying to find my proper niche in Canada.

That short sojourn dispensing booze to the thirsty population of Port Arthur and its surroundings did have its lighter moments. Drunks tried to get a bottle replaced for free by persuading the manager that the original bottle had slipped out of their hands while on the way home—the rule being that if you broke a bottle within the store it would be replaced for free but beyond the door you were out of luck. Then there was the wild-eyed, noisy French Canadian who demanded a bottle of "focking whisky," much to our consternation. We were ready to eject him from the premises when the manager arrived and supplied the man with his Focking Whisky. It turned out that it was a legitimate brand of Scandinavian whisky, and that the man deliberately made a great song and dance about it to embarrass us and perhaps get a bottle for free.

Chapter Five

Ashore and Afloat

It was now 1951 and I was still adrift, but ominous events in Korea were overshadowing my own personal concerns. The Chinese counteroffensive had swept the Allied Armies back south across the 38th parallel and, in January, had taken Seoul and were pressing southward. Canada had responded to the UN call for help in Korea, and had raised and dispatched an infantry brigade and armoured support to join the Commonwealth Division fighting alongside the Americans and South Koreans. Additionally, the Royal Canadian Navy had sent three destroyers to Korean waters. There was a general wake-up to the Soviet threat, and additional infantry and armoured units were sent to Europe. These units were backed by a fighter-bomber group—all under NATO command.

I had determined that a complete anchor swallow was beyond me but, with the freeze-up on the Great Lakes, I had no choice for the time being but to seek any sort of employment to keep the wolf from the door. I found work once more at the Abitibi paper mill, as a labourer on the sorting table, where I sweated away pulling logs off the moving belt that had not been properly barked. I would then feed these logs back into the barking machines, being extremely careful not to get tangled up in the fast moving parts. It was hard physical work but quite satisfying. The other workers all seemed to be large Ukrainian fellows, lately arrived in Canada as displaced persons. They had little to say and were fueled by the sausage and black bread that they brought for their lunch break. Some of them had obviously had a very tough time of it in Europe before getting out of the DP Camps set up after WW II ended.

Below the sorting table and the barker machines was an extensive labyrinth of rooms and passageways where the bark was conveyed toward a tunnel that led to the power plant where the end product was used as fuel. When all went well, the wet bark was moved on rubber conveyor belts through various shredding and drying processes. Quite often, however, the bark would pile up between belts, thereby bringing the entire operation to a grinding halt. If one was working on the sorting table, the noise of the alarm siren meant a welcome break in the hard physical work. But sometimes, the chaos developing down below required all hands to help out, and what a messy, wet, smelly operation it could be. At night, it resembled an artist's depiction of what parts of hell must look like, the only things missing being the flames and the "Old Fellow" himself. Visualize dungeons filled with a mixture of bark and water, drains blocked, equipment covered in wet bark, workers stripped to the waist pitch-forking wet bark in every direction, others diving into the cold water attempting to unplug the blocked drains, all the time being supervised by the little Slovakian foreman screaming at us in a mixture of European languages and good, solid-sounding cuss words. There were times that I was close to collapsing in hysterical laughter.

The pay was reasonable and I was quite happy to stay at the paper mill, but we were the extra shift hired to take care of surplus logs. So, after about a month, my Ukrainian comrades and I received our pink slips. The following day, I returned to the shipyard as fitter's helper. It was now February and working in the yard was a very cold experience, particularly after the indoor atmosphere of the paper mill. I learned to seek out the riveters and their braziers, or find an excuse to visit the unheated assembly shops, where at least one was out of the bitter wind. I longed for the spring when I hoped to ship out again on an upper laker.

Meanwhile, I got to know the city of Port Arthur and its people a bit better. I enjoyed two watering holes—the Arthur Hotel downtown and the local branch of the Royal Canadian Legion, which was located close by. The Arthur was handy to a bus transfer point, so got most of my business on working days, while the Legion was for relaxing during the weekends. The Arthur was a typical Ontario beer parlour, but with a unique feature—a large carving lying on the mantelpiece over the fireplace. It was a globe depicting the British Empire, with a huge lion on top and the words "What We Have—We hold!" The owners of the Arthur were the Arthur family, originally from Ulster, so the sentiments were understandable. The Legion was a more modern building, with facilities for entertaining and even contemplation. It was decorated with some unique paintings of battle scenes that the Lake Superior Regiment had taken part in over the years. There was an early one of the Riel Rebellion of 1885, then a vivid scene from the Battle of Paardeburg in the Boer War, followed by scenes from Vimy and Ypres in the Great War and, finally, by spectacular depictions of the actions in which that the regiment participated in northwest Europe during World War II.

My Legion friends were still trying to settle down in jobs of various kinds after their return from service. Among them were a city administrator, a lawyer, a miner, a funeral director, a speculator, the harbour master, and sundry others in more lowly occupations. We were united by the past and by our thirst. The Legion was also a good place to get out of the cold weather. My wife's other sister had also married a local veteran, Red Andrews, who had his own small trucking business. He used to join Bobbie, my father-in-law, and me at the Legion for games of pool or billiards. All in all, our gang kept that place going. I was hired from time to time as a spare barman or waiter when dances or other lively events took place. At these affairs, beer consumption soared and tension sometimes exploded into anger and physical confrontation. I often found myself attempting to pacify irate loggers or miners while being insulted and thumped for my pains. One night, I finally lost my cool when a big logger called me a "f---ing Polack!" I went over the table at him and had to be subdued by my friends. Why I reacted so strongly to being called a Polack, I cannot really explain. I actually had much admiration for those I had met during WW II. I suppose it must have been the contemptuous look on the miner's face as he detected my foreign accent. It was a grand place, the Royal Canadian Legion in Port Arthur!

Figure 10 - SS *Vandoc*, 1951

There was a large Great Lakes shipping company based in Fort William, so in early April I made myself known to them and, after being interviewed by the marine superintendent, was engaged to join the SS *Vandoc* in mid-April. I had to proceed by train to Collingwood on Georgian Bay where the vessel was refitting. She was an older upper laker of 4488 gross tonnage, propelled by the usual triple-expansion steam engine. The Master was a Captain Norris, a rather nondescript individual who dressed in his oldest clothes and generally exuded scruffiness. The mate was a Georgian Bay farmer-type of similar aspect. The second mate was a more colourful man, being of Italian background and of an excitable nature. The helmsmen and the deckhands were Georgian Bay–born and bred. I was made to feel an outsider, with my deep-sea connections viewed with a great deal of suspicion. The second mate became my only ally on deck. The rest of the crew, engine room, and galley were largely pleasant and companionable.

Vandoc's initial voyages of the season were to the Lakehead to load grain for the lower lakes' elevator terminals. In mid-May, we switched to the iron ore trade, loading at the ore docks in Port Arthur and delivering our cargoes to steel plants along the US shore of Lake Erie. It was not exciting work, but left little time for relaxation.

The Legion in Port Arthur seemed to be a long way away. On deck, I was given little opportunity to practice my navigational skills and otherwise was made to feel like a useless appendage. Additionally, I was ostracized because I would not take part in the floating poker game on board, a game manipulated by the mate to rob the young sailors of their hard-won pay cheques. It was a sad affair that reflected badly on the ship. I knew that I could be difficult to get along with at times, and that perhaps I was too outspoken for my own good, but this was not a happy ship. Just at this point my world changed, this time for the better. My wife contacted me regarding a message received from the Canadian Merchant Service Guild offering me a position as second mate on a deep-sea freighter. I immediately told her to cable my interest and began to take steps to leave *Vandoc*. All my funds were in Port Arthur and I needed money to travel to Montreal. The second mate very kindly lent me enough money for my travel, and I departed the vessel in Windsor, Ontario, on the 3rd June.

At that time in Korea, the Chinese had been pushed out of Seoul and were retreating back to the 38th parallel, but all I could think of was getting back to the deep sea. I reported in Montreal to the offices of Lunham & Moore Shipping Ltd., the owners of the SS *Angusglen*, my next ship.

In writing this epistle, I try to convey my own experiences in relation to the local or broader context. It occurs to me that I have waffled on at some length about my voyages on various Great Lakes vessels without ever attempting to describe the actual Great Lakes of North America. They consist of five lakes—Superior, Michigan, Huron, Erie, and Ontario—and are the largest system of fresh surface water on the face of the Earth, eighteen percent of the earth's supply, with only the polar ice caps containing more water. The Great Lakes contain 23,000 cubic kilometres of water spread over an area of 244,000 square kilometres that stretches 1,200 kilometres from west to east. The Lakes, with their connecting waterways and tributaries, became the avenues of penetration for the European explorers and settlers. It started with La Salle who explored the lower lakes in the seventeenth century with the help of local Indian tribes using their main means of transportation—the canoe.

Settlers followed, utilizing small sailing craft. In the eighteenth century the struggle for domination of eastern North America began between the French based in Quebec and the British based in their eastern seaboard colonies. Eventually the British expanded the war to subdue the French fortress at Louisbourg on Cape Breton Island, which controlled access to the St. Lawrence River. This action was followed shortly thereafter by the fall of Quebec at the Battle of the Plains of Abraham, which marked the end of French power in the Great Lakes and on the eastern seaboard.

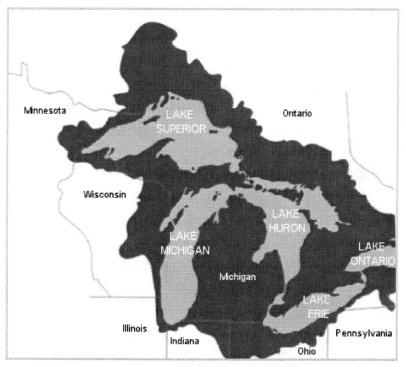

Figure 11 - The Great Lakes

Expansion of settlement and trade on the lower lakes proceeded slowly, being delayed by the American War of Independence (1775–1783). The end of that war brought about a flood of settlement to the south shore of Lakes Erie and Ontario, and the arrival of many United Empire Loyalists who had lost everything and who settled along the north shore of the St. Lawrence River between Kingston and Montreal. The peace lasted for almost thirty years and was then broken by the War of 1812, when US and British Men-of-War fought for control of the Great Lakes. The battles took place on land and sea, with the north shore of the St. Lawrence River remaining in British/Canadian hands, together with the Niagara Peninsula. The British lost their outposts at Detroit and Sault Ste. Marie. Peace brought great prosperity and expansion all around the Lakes.

The first steam-propelled vessel on the Great Lakes was built at Kingston, Ontario, in 1816. She was the *Frontenac*, of seven hundred net tons and propelled by a 40-HP engine. Her captain was a Scotsman, James McKenzie, and her engineer was another Scot, John Leys. They loathed one another, so it was a wonder that the ship survived without too much incident until she was broken up in 1827.

In the nineteenth century, there was an enormous expansion of settlement, followed by trade and commerce. Passenger steamers, bulk carriers, tugs and barges, fishing vessels, and others made the Great Lakes a very busy highway indeed. It was also an age when many of them foundered in the stormy seas, collided, exploded, or were swept by fire. A roll call makes sobering reading. The canals, originally built by the British for military purposes at Lachine/Prescott, Ottawa/Kingston, Trent/Severn, Welland and Sault Ste. Marie, all played their part in the fast growth of the region.

The expansion continued into the twentieth century, with bulk carriers replacing many of the passenger steamers that had lost their battle with the railroads. In 1949, the last big passenger steamer, *Noronic*, perished by fire along with many of her passengers while docked in Toronto. An era was at an end. Now the day of the self-unloading bulk carrier was at hand, as the last of the package carriers were laid up. The transport of iron ore, grain, and coal was the lifeblood of the carrier fleet. In 1951, my last year on the Great Lakes, the concentration of vessels of all shapes and sizes at choke points such as the Sault locks, the Strait of Mackinac, the Lake St. Clair/Detroit waterway, and the Welland Canal were sights to behold.

In 1951, thirty million people lived in the Great Lakes basin. The industrialization of a century and a half was taking its toll. Eutrophication of parts of Lake Erie was well advanced, and concentrations of toxic wastes were harming various areas in Lakes Ontario, Erie, and Michigan. The conurban centers of Toronto/Niagara/Buffalo, Detroit/Windsor, and the Chicago area were bringing death from industrial pollution to many people. It took a massive joint

Canadian/US management effort from the 1960s onward to control and eventually reverse these dangerous trends.

One final note: The completion of the St. Lawrence Seaway in the late fifties enabled large deep-sea ships to access the Great Lakes. Today, much of the Great Lakes traffic is of a foreign-going nature, with Canadian- and US-owned ships, plus those of other nations, taking their cargoes directly to their final destinations, thereby surmounting the old problem of cargo transfer around the St. Lawrence River rapids. All of this activity goes on without the mass of the population, concerned with its own affairs, being aware of the importance of this seaborne traffic to their own ultimate well being.

Now on to the *Angusglen* and the deep ocean.

Chapter Six

Return to the Ocean

My time in Montreal was limited to a few days, being interviewed and selected for the position of second mate on the SS *Angusglen*, touching base with relatives and preparing to fly to Newfoundland, where the vessel was located. The last night in Montreal, I celebrated by visiting several clubs on St. Catherine Street, drank too much, and almost got into a serious altercation with a French Canadian who was even more looped than me. However, early the following morning, I was boarding an aircraft at Dorval bound for St. John's. It was a bumpy flight and I suffered accordingly. Also on board was an older chap called Wilson who, it turned out, was the new first mate recruited for *Angusglen*.

Figure 12 - SS *Angusglen*, 1951/1952

At the St. John's airport, we were met by the ship's agent, who took us to Portugal Cove on Conception Bay, where we boarded a small motor launch that took us across the water to Bell Island, where our vessel was preparing to load a cargo of iron ore bound for Scotland. The weather was grand that evening and the thought of returning to my native land was a wonderful bonus. *Angusglen* was a typical wartime-built freighter of around seven thousand gross tonnage with a triple-expansion engine and a cruising speed of ten to eleven knots. She had formerly been named *Fort La Have*, having been constructed on the west coast of

Canada to British Ministry of War Transport specifications. She had been trading between North America, the Canary Islands, and Spain, and was now contracted for a series of voyages from Newfoundland to the UK with cargoes of iron ore.

It was a pleasure to settle down on board a real deep-sea ship for a change, and indeed I was made very welcome. The captain was a chap called Smith, a youngish man originally from the Shetland Islands. The second mate I was replacing was also a Scot, returning to Scotland as passenger on *Angusglen* to sit for his Master Foreign-Going Certificate. The third mate was Greg Bellavance from the Rimouski area of Quebec. The chief engineer was named Orr, based in Halifax but clearly an Ulsterman by birth. The second engineer was also Irish but located in Montreal. The third engineer, an Englishman, was a familiar face from my days on *Peribonka*. The fourth engineer was a most unusual man named Hollis who had been an outstanding soldier in World War II and had been awarded the Victoria Cross. The fifth engineer was a wild young fellow named Weir from Montreal whose alcoholism soon led him astray. The radio operator was an older man who had spent much of his youth in South Africa. The chief steward was a Newfie who fancied himself a ladies' man. His trying to keep us reasonably well fed was balanced against his desire to seek profit for himself.

The foregoing were all housed midships on *Angusglen*, as were the bosun, the carpenter, and the donkeyman. The rest of the crew were housed in the ship's stern accommodation. They were a mix of Montrealers, Maritimers, and Newfoundlanders, all knowledgeable about the sea and with an opinion on everything, as I soon discovered. She was a happy ship, that much I already knew.

Where we lay alongside was known as the port of Wabana, the iron ore loading facility that lay beneath the eastern cliffs of Bell Island, whence the iron ore was mined from deep underground. The cliffs were about three hundred feet in height, thereby providing a natural access for the iron ore chutes poised over the ship's holds. Bell Island could only be reached by climbing up almost vertically—several hundred steep steps. Only the fit need apply!

We departed Wabana on the 6th June 1951, bound for Glasgow with a full cargo of iron ore. The weather was perfect, little wind and no sea to speak of, with a gentle northeasterly swell—unbelievable conditions for the North Atlantic. Six days later, we were approaching the Cumbraes in the Firth of Clyde, where we picked up our pilot. The weather was still almost perfect, with the highlands to the north standing out against the evening sky and the Renfrewshire coastline slightly obscured by a Scotch mist that often hangs over Gourock and Greenock, even on the best of days. We proceeded beyond the Tail of the Bank, up river to

Rothesay Dock in Clydebank, our port of discharge. Unloading commenced at a leisurely pace, with overhead cranes lifting our cargo by clamshell into the small railway wagons lined up along the dock. It was obviously going to be a long affair.

My father and stepmother still lived in Greenock, so I took an early opportunity to visit with them. On the train going to Greenock, I met an old shipmate of mine from our days in Egypt awaiting repatriation in 1943. He was one of the survivors from the torpedoing of *Yoma* off Derna, one of two engineers who were lucky to get out of that engine room in time. The meeting led to a visit to a local pub, which ensured that I was slightly squiffed by the time I got home to Dad. My timing was not good, and my father was not pleased.

However, I was able to make several visits home, and gradually repaired fences with my father. My stepmother of course supported Dad, and gave me a decidedly cool reception. I surely did know how to upset people! Meanwhile *Angusglen* continued discharging our cargo at an extremely slow pace, so there were lots of opportunities to visit Glasgow and take part in its culture and low life, as one desired. A visit to the Kelvingrove Art Gallery and Museum could be followed by a raucous attendance at a football battle at Ibrox Park, which led to a noisy celebration in one of Glasgow's many noisy pubs. It was as if I had to soak up all the local atmosphere before being wafted away, back to the harder, more basic sort of life I had been living in Canada.

The war in Korea and the Soviet threat worldwide were having an effect on the economy of Clydeside. The shipyards were booming, and the ports were busy with vessels and their trade. The memories of the war were fading slowly, with some items still in short supply. Meat was a particular problem with Argentina, Britain's long time supplier, limiting severely its exports of beef to the UK. The dead hand of socialism permeated the land. That, above all, made sure that even though I loved being home on a visit, there was no way I could contemplate a permanent return. My fate would be sealed in Canada.

After almost two weeks in port, we completed discharging and, after refuelling at Old Kilpatrick, headed out to sea. The weather was moderate in the North Atlantic and so we made good time to Conception Bay. I was in my element, being given full rein by the Master to control the navigation of the vessel. In addition to the usual sights from land objects, I took many star and sun observations every day of good visibility. Updating the nautical charts from notices to mariners, winding the chronometers daily, and tending the gyrocompass were all part of my normal routine. Recurring nightmares always included neglecting to wind the chronometers or start up the gyro in sufficient time for it to settle down properly for a sea voyage. That dream stayed with me all of my days at sea.

There was an additional interest on *Angusglen's* voyages. She was fitted with six passenger berths abaft the master's quarters under the bridge deck. All berths were occupied leaving Scotland with male and female passengers emigrating or returning to Canada. Their presence on board made watch keeping more interesting, and their stories were often fascinating. On this voyage, I particularly remember one man who had been a paratrooper in the British Army in Burma. He had been part of the British invasion force that recaptured Rangoon in the early part of 1945. As that part of the world held nostalgic memories for me, I soaked up his stories with enthusiasm, and smiled at his reminiscences of that exotic tropical land. However, all had not been plain sailing for him when he was dropped over Burma. He found himself landed about ten miles from Rangoon and on the wrong side of the river. Several days passed before he and his companions were able to stagger out of the jungle and seek boat transport across the river. By that time they were short of rations and reliable drinking water, and suffering from the depredations of myriads of insects of the jungle. Luckily, they did not run into any Japanese patrols, although their presence was detected.

We loaded our second iron ore cargo at Wabana in early July and were off, out into the North Atlantic in record time. The weather was still favourable, although moderate winds and seas were encountered from time to time. We had on board a few passengers returning to the UK, mostly female, among them the wife and child of the manager of the Hotel Newfoundland in St. John's. She was English but could tell wonderful tales of cod jigging in a genuine Newfie accent.

In no time at all, we were steaming up the Clyde to our berth in Rothesay Dock. I had a very nice visit with my father, who took me down to Largs, where we enjoyed the show at the theatre and loudly and passionately sang "A Gordon for me!" along with the entertainers and the audience. I also met up with my Aunt Bessie in Glasgow and, in a conversation with her over lunch at Guy's Restaurant, decided to seek Captain Smith's help in obtaining passage for Doreen on the next voyage to the UK.

During our sojourn in port, I was able to get to know several of my shipmates. Tommy Orr, the chief engineer, was a nice fellow who seemed to enjoy my company. We both liked good food and wine, and so gravitated to the more expensive and exclusive sort of establishments. A favourite place was the bar and grill at the Grand Central Hotel. Orr had served in armed merchant cruisers with the Royal Navy in World War II, mainly patrolling the area between

Shetlands and the Norwegian coast, on the lookout for German raiders. He later transferred to the Royal Canadian Navy before returning to merchant ships at the end of the war. He lived in Halifax with his wife. Their only son had just entered the Canadian Services College Royal Roads in British Columbia. Orr was an Ulsterman of the very best type—knowledgeable about the world, fair minded in his treatment of subordinates, thoroughly grounded in his craft, and loyal to his country and the ties that bind. I liked him enormously.

Another shipmate was the fourth engineer Stan Hollis. He was of rangy build, reminding one of Gary Cooper, except that his face was scarred with shrapnel burns, his upper body decorated with many skin grafts, and one leg reconstituted shorter than the other. He had had what was called "a good war," but had paid quite a price.

Figure 13 - Stan Hollis, VC, Junior Engineer, with radio officer, *Angusglen*, 1951/1952

He was a Yorkshire man who grew up working in the racing stables of northern England. He joined the Territorial Army in the late thirties, serving in the Green Howards. He saw some action in France at the time of the Dunkirk evacuation, before being shipped out to Egypt to join the 50th Division of the British Eighth Army. He was co-opted into a commando unit that did much of its work behind enemy lines. There, he received the bullet wounds that eventually shortened his leg. He respected his former German foes, but tended to regard his former Italian enemies with scorn and contempt. He recalled how the British Army would feel out the enemy lines to locate Italian divisions, and then attack them in full force to isolate them from their German allies prior to destroying them or taking them prisoner. He also described hiding in the sea surf by day and marching at night when caught behind the lines, sometimes getting help from Libyan tribesmen who would hide them among their flocks of sheep.

He was decorated several times, but his greatest award came later in the war when he received the Victoria Cross for his successful capture of German gun batteries on the left flank of the D-Day landings in Normandy. He was the only unwounded survivor of his company of commandos. He went on to lead his reinforced unit in further battles in Normandy and beyond. Company Sergeant Major Stan Hollis was indeed a genuine hero. His final responsibility prior to being demobbed was as commander of the Honour Guard after the liberation of Paris, where his difficult job was to ensure his and his men's sobriety for the daily flag ceremony. He was not an easy man to get to know, but he did like to socialize, particularly over a pint or two. There will be much more to the Stan Hollis story.

Captain Smith advised me that a berth would be available on *Angusglen* for Doreen as passenger on the next crossing from Wabana to the UK, so I contacted her by cable to make the necessary arrangements to travel from Port Arthur to Newfoundland. Upon completion of unloading in Rothesay Dock we headed back to sea. The voyage across the North Atlantic was uneventful and we sailed into Conception Bay in early August, to be confronted by numerous vessels lying at anchor. The mine workers had gone on strike!

The strike lasted for several days, causing me some consternation as I tried to contact my wife. Eventually, she was located at the Hotel Newfoundland in St. John's, awaiting instructions. Meanwhile, the mate and I and several others managed to get ashore on Bell Island in a leaky old dory, and patronized the local Legion until closing time. No boat could be found to take us back to the ship, so we found a home where the family put us up for the night. Two to a very narrow bed did not ensure comfort, but it was better than wandering around all night in the dark. We found a willing dory owner early the following morning and arrived back on board, but not before some callous idiot flushed the midships toilet over the top of us. A fitting end to an ill-conceived adventure.

The following day my elegantly dressed wife arrived on board and immediately commented upon my scruffy attire. That hurt, but it was true, I was no longer the well-tailored young officer she had married. The years on the Great Lakes had taken their toll. I smartened up as much as I could and gave her all the attention that she deserved.

Figure 14 - My own second mate, *Angusglen*, 1951/1952

She had a tale of her own to tell, making her way from Port Arthur by train and ferry to Conception Bay. My wife had a small but expensive wardrobe of stylish clothing, and she was very pretty, so she received a great deal of male attention en route. Upon arrival in Sydney, Nova Scotia, she had to await the arrival of the ferry to Port-aux-Basques. She could not find a restaurant or hotel open, so she was directed to a large house where she was met by an elderly woman, who seemed puzzled by her appearance. However, she was invited indoors to partake of a cup of tea, and allowed to rest on top of a bed in a bedroom, as long as she did not rumple the bedclothes. Later, several other younger women appeared, who were fascinated by her smartly designed suit. She was puzzled by her reception, and it was only later, when she left the house to join the ferry, that she discovered she had been entertained in a house of ill repute. Her clothing and foreign accent, together with her good looks, had led people to believe that she was a new recruit. The ferry trip was rough but otherwise uneventful. She then boarded the famous Newfie Bullet, which rattled across the breadth of Newfoundland to deposit her in St. John's. On the journey, she was entertained by a group of Englishmen employed by Bowater, the pulp mill and paper production company. It was difficult to find suitable accommodation in St. John's, but the shipping agents persuaded the manager of the Hotel Newfoundland to put her up in his daughter's bedroom, she being absent on a trip to the UK. What a tale she could tell to her children and grandchildren.

The strike was settled and we proceeded alongside to prepare for loading. Advantage was taken of the lull prior to loading to explore Bell Island. With Doreen in tow, we climbed the three hundred and sixty-five steps to the top of the cliff and made our way to the drinking establishments of Bell Island. Stan Hollis led the parade and we soon landed again at the Canadian Legion. When the locals found that they had a genuine Victoria Cross winner in their midst, the drinks really flowed. Before closing time, Stan was presented with many mementoes, including bottles of Scotch whisky and gin. When we reached the top of the steps leading down to the ship, Stan gave Doreen the responsibility of looking after the liquor as it was transported down these many steps. Why he did this I will never know, as she was somewhat inebriated herself. But God often looks after fools and drunks, and she successfully navigated the stairway and then climbed the steep ladder to board *Angusglen*. Mission accomplished!

We departed Wabana fully loaded the following afternoon, with a number of people who will be nameless suffering from gigantic hangovers. Our run across the North Atlantic was mostly uneventful, but it was grand to have my wife on board. She soon adjusted to shipboard life and, as the only passenger on board, had the run of the vessel. Much to her delight, we were ordered to Merseyside to unload our cargo. It was late August when we docked, and Doreen's relatives gave us a truly warm Scouse welcome.

During our stay in Liverpool, I played a trick on my wife that nearly backfired badly. A group of us had gone with Doreen to the pub that was situated just outside the dock gates. She carried no identification, so she was stopped upon re-entry at the dock gates. The rest of us were asked to identify her and I jokingly indicated that she was just someone we had picked up in the pub. The joke was accepted at face value by the dock policeman on duty, and she was denied re-entry to where our vessel lay. A very annoyed wife stewed at the dock gates while I hastily made my way on board to seek the Master's written confirmation of her legitimate presence on the ship. Needless to say, I was in the doghouse for some time.

After we had completed discharging our cargo of iron or, we proceeded to dry dock for minor shaft repairs. Dry-docking was very much a manual affair orchestrated by bowler-hatted chaps bellowing instructions over megaphones to a mob of workers manhandling our hawsers. A similar operation in North America would be conducted quietly and with fewer personnel. At this point in time, a number of our crew who were horseracing fans slipped away to witness the famed St. Leger race, while no doubt losing their hard-earned cash.

I left my beloved in the care of her Aunt Harriet while *Angusglen* departed the Mersey for Newfoundland. We had a good passage across the North Atlantic, with a full complement of passengers on board. One interesting fellow was a German ex-U-boat officer who told me emphatically that Germany did not lose the Battle of the Atlantic by being beaten by the Allies, but by our replacing large losses of merchant ships with enormous quantities of replacement vessels. There was an element of truth in his statement, even if arrogantly expressed. Another interesting passenger was a professor from Memorial University of Newfoundland whose expertise was in the study of whales. Whenever we spotted one, he practically lived up on the bridge viewing its appearances and reappearances. He was very concerned about the dwindling population of these large animals throughout the world.

We docked in Wabana in late September and changed Masters. Captain Smith had been a good skipper, all on board agreed, and we all wished him well. Our new Master was Captain Oliver Bray, a taciturn Australian who had seen wartime service in the Royal Australian Navy during the battles with the Japanese in the Solomon Islands and in the Coral Sea. We soon discovered also that he was deeply religious and was rumoured to be a follower of the famous faith healer of the day, Father Divine. However, none of the foregoing spilled over into the day-to-day operation of the vessel, and our routine activities remained unaffected.

Our fourth load of iron ore was taken on without incident and we departed Wabana for the UK, experiencing moderate weather conditions for most of the voyage. We were back to our original port of call at Rothesay Dock in Clydebank. Doreen traveled up from Liverpool to be with me, and we spent precious time with each other and in visiting Dad and my stepmother. Through my father, I kept up to date with the exploits of my brother Gordon, who appeared to be surviving the stresses and strains of the Communist terrorism campaign in Malaya. It was a nasty fight aimed mainly at the plantation industry and the tin mining operations. Managers and their wives and children were all prime targets, and no one knew how it would all end.

In October, *Angusglen* returned to Wabana for her fifth, and what turned out to be, her final cargo of iron ore destined for the UK. The weather in the North Atlantic was now becoming stormy and the passage increasingly uncomfortable. We loaded our cargo at Wabana as usual, but this time we loaded in error beyond our normal loaded draught. It was an apparent windfall for the company for which we were to pay dearly. We returned to Rothesay Dock for discharging our cargo but grounded while berthing. Those extra tons and deeper draft had made a difference. We had inflicted damage upon ourselves, but were as yet unaware of the extent of the problem.

Doreen joined *Angusglen* in the Clyde in preparation for the voyage home. Several additional passengers joined the vessel, and we made our way in ballast out through the North Channel into the North Atlantic. It was sad to say farewell to my father, and I hoped that it would not be too long before I saw him again. The weather was foul during all of our passage and deteriorated even more as we passed the mid-point of our voyage. Then, potential disaster struck when the ship's carpenter detected a large amount of water swilling around in Number 2 hold. It took the combined efforts of the crew to dive into the dirty water and, after finding a small crack in the hull close to the bilge keel, to temporarily place a wooden wedge in the crack and pump out most of the water in the hold. At the height of this crisis, *Angusglen* was only thirty miles or so from the entrance to Conception Bay, but a full gale developed that blew our ship many miles to the northeast. It was twenty-four hours before we were safely inside St. John's harbour where we had run for safety.

In such miserable weather, all our passengers were suffering from seasickness, including my wife, who suffered more than most. In her pitiful state, she requested that her husband find her a dainty chicken sandwich to settle her stomach. Instead, the unspeakable clod presented her with a sandwich dripping with mayonnaise and lettuce, which she was unable to eat. He then proceeded to demolish the large sandwich while she writhed in extreme distress on the bunk!

Once secured at the wharf in St. John's, an immediate call went out for an inspection of the damage to our ship's hull, and we entered dry dock without delay. The crack on the bilge keel plate was clearly exposed. It ran about nine inches long in fore and aft direction and would need plate replacement before we were pronounced fit for our normal seagoing trade. However, we were given permission to proceed to Halifax, providing we completed a temporary repair with a substantial cement box. This temporary replacement was completed under my personal supervision working over the weekend, and received the prompt certification of inspection required.

Angusglen then proceeded southward to the port of Halifax, a distance of roughly five hundred miles, in slight seas and wind with moderate swell. Meanwhile, my wife's disposition had improved remarkably and she was able to regale me with some choice stories of the antics of her fellow passengers. They were apparently a lively lot, but the most interesting was a Nursing Sister who was returning from leave in England, en route to the Grenfell Mission at St. Anthony close to the northern end of the Strait of Belle Isle. She was quiet and demure on the surface, but apparently concealed a passion for our captain. There were several nocturnal visits to the captain's cabin, ostensibly to use his bathtub, and other contacts between the two of them that raised speculations considerably. As they were both of a fundamentalist religious bent, they were probably suited for one another. There were tears on the wharf as they took leave of one another, but if there was a happy ending to the story, we were unable to confirm. A quiet man was Captain Oliver Bray.

When we docked in Halifax, it was early December 1951, and the weather was mild and welcoming. A pre-Christmas spirit permeated the port, and we were made very welcome. Several parties were arranged, and much liquor and seafood were consumed. Oysters and lobster became our main fare, with Doreen developing a passion for the crustacean of the North Atlantic.

It was a sad time too, as several of our shipmates signed off and departed for their homes. However, many signed on for the new voyage, together with me, as we completed repairs and prepared for sea. The weather turned much colder suddenly, and several of our steam winches seized up in the cold snap.

Finally, the day arrived to say farewell to my wife, who was returning to Port Arthur. She had been a good shipmate, and would be sorely missed, not just by me but also by those who shared accommodation amidships on *Angusglen*. She had lived among us and knew only too well that none of us were saints—but she had influenced our behaviour significantly and imparted a sense of female sensibility to the sometimes rough and harsh seagoing atmosphere in which we existed. It was a sad farewell as she boarded the train for Montreal and Port Arthur.

The ship departed Halifax on the 20th December, bound in ballast for Norfolk, Virginia, to load a cargo of coal.

Chapter Seven

To the River Plate and Rio Grande Do Sul

Running in ballast south to the entrance to Chesapeake Bay was not exactly pleasant in late December, but we made good time and were soon anchored in Hampton Roads awaiting the availability of a coaling berth in Norfolk, Virginia. The roads were filled with ships, so we were obviously destined to remain at anchor for a few days. A few of us took a tender for a run ashore to Norfolk to see the sights. Stan Hollis, the radio operator, and I visited several nightspots, but eventually tired of it all and tried to return to *Angusglen*. However, the tenders appeared to have packed up for the night, and we were forced to seek temporary accommodation ashore. We found the Monticello Hotel and tried to obtain rooms but, as we could only proffer Canadian currency, we were unable to persuade the desk clerk of our bona fide intentions. However, he did allow us to sit in the lounge overnight, which we did until dawn came and we could seek marine transport. It was an uncomfortable night and, with hundreds of Canadian dollars in our pockets, we felt unduly snubbed.

We finally proceeded alongside the coaling wharves and loaded our cargo in a few short hours. An Italian ship had berthed astern of us, and her friendly crew were intent on doing business with us. In short order, we were the possessors of a number of large flagons of Chianti, which were polished off rather quickly. It was not particularly good wine but came at a very reasonable price. We later discovered that the wine had been broached from their cargo, but by that time we were off to sea.

We laid off a course for the Mona Passage, to clear Hispaniola and Puerto Rico on our way to Port of Spain, Trinidad, where we would refuel for the voyage down the east coast of South America to Buenos Aires. It was a largely uneventful passage, with pleasant tropical weather becoming the norm. We took on bunker oil from a barge while lying at anchor off Port of Spain. No shore leave was granted, but we did manage to purchase a case of excellent rum from one of the bumboats circling the ship. The rum lasted until we approached the River Plate, so it must have been a fair-sized case, even if shared with the chief steward and the radio operator.

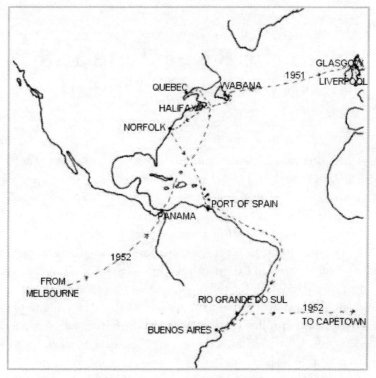

Figure 15 - *Angusglen* voyage around the Globe, Part I, 1951/52

The radio operator had a fund of stories about his days on the railway in Rhodesia, where being eaten by hungry lions, cheetahs, or leopards was a distinct possibility. Additional threats to life and limb came from rampaging elephants or water buffalo. According to him, the sound of the train seemed to infuriate many of the animals, and leaving the safety of the train at night to switch points or conduct any other activity was a distinctly hazardous undertaking. It all sounded adventurous and a million miles away.

The chief steward was a Newfie with a bag of stories about serving on the Cunard liners running to New York. They all underlined his sexual prowess among the female passengers and staff.

There was great anticipation on board as we entered the River Plate and took on the pilot for the tortuous approaches through the mud banks and shoals in the approaches to the port of Buenos Aires. The city was famed for its nightlife, which did not get into high gear until midnight, so the crew were all primped and ready shortly after we secured to the wharf in the early evening. The choice of places to go was wide and ranged from the high-class but expensive bars in the vicinity of the Avenida Florida to much wilder establishments in the dockland Boca or Barracas zones. The sensuous sound of the tango filled the air, with a beat of Latin drums seeming to come from every doorway and corner. The ladies were available everywhere, liquor flowed, and the air was filled with the smell and sizzle of the grilling of good Argentinean beef. Grand fun, but we had to face a 6:00 A.M. unloading start in the morning, which curbed our appetites more than somewhat.

Figure 16 - My usual working rig

It truly was an exciting city to visit, even though Juan Perón and his wife Evita controlled the country through a fascist-type dictatorship. Indeed, everywhere one went, huge posters confronted one declaring "Perón cumple, Evita dignifica." My lack of knowledge of the Spanish language did not prevent me from understanding the propaganda. All the many parks and public buildings were festooned with these posters depicting their glory. It was also soon apparent that, although the language was Spanish, more than half the population appeared to be of Italian background. Even as we lay in port, large Italian liners such as the old *Saturnia* were discharging thousands of Italian immigrants into the country. Their influence could be seen everywhere, but particularly on the poor quarters adjoining the port such as the Boca and the Barracas.

The Argentineans were passionate about many things, but gave much of their attention to football and to the racetrack. The Boca Juniors and the River Plate were teams with a rivalry similar to the Rangers and Celtic in Scotland, and dominated the game. Horse racing was almost a daily occurrence at the many tracks surrounding the city, such as Palermo and San Isidro. I am not really a betting man, but followed expert Stan Hollis to the betting wickets. There I learned that "Ganador" meant "On the nose," and that Perón supporters rejoiced in their nickname "los descamisados" (the shirtless ones). We also visited a nightclub called "Aldeanos" that was notorious during the war years as a hangout for German spies. It was therefore off limits to all British personnel. It was a lively spot when I was there, with a lederhosen-dressed brass band and a gigantic bell that one had to be careful not to ring, as it indicated that the ringer was prepared to pay for a round of drinks for everyone present.

The country appeared fairly prosperous, but was dependent on its exports of chilled or frozen meat to Europe. It was also a large exporter of grain. The beef and the grain were grown on the pampas, the vast well-watered plains stretching westward and northward from Buenos Aires. Many of the large cattle ranches were still owned by wealthy British families, who also had had enormous interests in the railways that brought their cattle to market. These railways had long since been nationalized but still retained a British flavour through their retention of many of the original managerial staff. The influence of the British was still to be seen in Buenos Aires in the private clubs, horse racing, and polo playing. As an illustration of their presence, I attended a cocktail party on board the *Highland Brigade*, a Royal Mail Line passenger vessel lying farther along the dock. This visit was sparked by my friendship with her second mate. The party was very convivial, and in no time I was becoming acquainted with a bevy of local belles with the non-Latin monikers of Barrington-Smith, Philips-Cargill, and a captivating one named Lillian Huggard-Caine. It was a memorable experience for a rough-hewn mariner.

We eventually completed discharging our cargo destined for Buenos Aires, and departed up the Paraná River bound for Villa Constitucion, a small port close to Rosario, the major city located on the river. It was now February 1952, the weather was sticky and hot away from the delta of the River Plate, and the river water was coloured a muddy brown. The startling green grass of the pampas stretched on either side of the river to the far horizon. Many fair-sized fish lived in the muddy river. They made good eating were but rather disconcerting to view as they were caught, being covered with spikes and antennae to enable them to navigate and survive in these waters. We pulled alongside the wharf at Villa Constitucion, which turned out to be a coaling supply port for the railroad. It was really a very small town, with a typical small square. The square was deserted most of the day, but livened up in the early evening, when courting couples with their guardians promenaded in their finery. It was also an opportunity for the unescorted batches of young men and girls to get a good view of the talent. Very old fashioned, no doubt, but rather nice to watch.

There now occurred one of those events that can change one's thinking for all of time. The news was received by shortwave radio that King George VI had died. The sad news was passed on to the crew, most of whom expressed little interest, and some made most disrespectful remarks, even objecting to the lowering of our Red Ensign. I had never been a strong monarchist, but retained a healthy respect for the institution. This blatant disregard of our history and antagonism to the monarchy was new to me and made me realize that there was a big unresolved problem of Canadian identity. I knew which side I stood upon.

The forgoing situation was brought into deeper focus by the discovery that several ex-Nazi SS men were living in the area close to town. Stan Hollis, who spoke some German, had located them in a bar fairly close to the docks. They were young, blond, blue-eyed supermen who made no attempt to hide their background, even boasting about it, much to Stan's anger. As a commando he could expect no mercy in war from the Waffen SS, and gave none in return! His stories of wartime actions took on new meaning. How had these men got to Argentina after the war, and why had Perón and his henchmen allowed them to remain? We have no solid answer to these questions even today.

Figure 17 - *Angusglen* awaiting rice at Rio Grande do Sul, 1952

Angusglen completed discharging her cargo of coal and moved down river from Villa Constitucion out into the River Plate estuary. Our next port of call was Montevideo, in Uruguay, where we stopped briefly to take on fresh supplies of food, particularly lamb, which was very reasonably priced. Our next destination was the port of Rio Grande in the state of Rio Grande do Sul in southern Brazil. Rio Grande was a moderately sized port city located at the entrance to a shallow water lagoon, Lagôa dos Patos, which ran more than a hundred and fifty miles northward to the provincial capital city of Pôrto Alegre.

Our cargo was to be rice that we would take to the East Indies, the rice bowl of the world. Insurrection in much of the East Indies and in Burma had created real shortages of rice, and it was our job to fulfil that need. The rice was delivered alongside *Angusglen* by sailing barge. These sailing barges came down the lake from Pôrto Alegre and were often at the mercy of adverse winds, thereby delaying our already pitifully slow loading progress.

We began to feel a part of the city and its activities. The cathedral was quite close to our berth, as was the small red-light district that seemed to rent its buildings from the Church. There was a casino of sorts that usually provided good food, although I broke a tooth there one night. I also had a substantial win, which I squandered in various ways before departing Rio Grande. Further along the wharf was a meat packing plant that seemed to specialize in producing corned beef. A grand tour of the establishment turned me into a temporary vegan. The plant employed a large number of women who soon made the acquaintance of our crew members by sauntering slowly by our vessel each day. We were instructed not to allow them on board, but it became a hopeless task, and one got used to seeing female garments strewn around the crew's quarters.

We were there at carnival time, and that was a sight worth seeing. Large numbers of people out on the street at night, lights everywhere, bands hammering out sambas, rumbas and other Latin music, scantily-dressed women spraying passers-by with coloured water—all designed to make you partake of the carnival spirit. Additionally, food stalls and refreshment tents were placed strategically to ensure that all enjoyed themselves. The show went on night after night for about a week before life returned to normal.

When we completed loading and ensured proper hold ventilation for our cargo, we anchored off to conduct a thorough fumigation of the vessel. I stayed on board to keep anchor watch while everyone else went ashore overnight to stay at hotels and rooming houses. The actual fumigation was conducted from a barge moored alongside and manned by several fair-skinned Brazilians who turned out to be of German extraction. Apparently there were many German settlements in the province, dating back to the middle of the nineteenth century.

With the ship back to normal the following day, we departed Rio Grande en route to Capetown, and then Surabaya in the former Dutch East Indies.

Chapter Eight

Eastward across the Oceans to the Orient and the Antipodes

Angusglen departed Rio Grande in late March 1952. A course was laid to roughly follow the 36th parallel of latitude south across the South Atlantic that would take us directly to Capetown, South Africa. The weather was mostly pleasant, with moderate seas and winds encountered along the way. The albatross of the southern oceans was ever present, dipping and wheeling over the trough of the waves and soaring high over peaks of breaking waves. A truly awe-inspiring sight and reassuring us that we were not alone.

Halfway on our journey, we passed close to the south of the island of Tristan da Cunha, a British possession struck with volcanic eruptions. The tiny population was later evacuated and relocated to the UK, but were so unhappy that they were returned later to their island. No vessels were seen en route until we approached the African coast.

It was time to reflect on the world in general. The situation in Korea was still tense, but an armistice was being negotiated at Panmunjom. Meanwhile, the Allies and the Chinese fought bitterly for strategic hills along the 38th Parallel. In Europe the forces of the North Atlantic Treaty Organization and those of Russia and its Communist allies faced one another, with Germany a particular flashpoint. The Russian build-up of its nuclear capability was of particular concern. British control of oilfields in Iran was under nationalistic threat. Independence was in the air, and several British colonies had become sovereign nations with the assistance of their former ruler. In others, such as Malaya, the situation was not so clear cut. The independence movement employed terrorist tactics and was of a particular ideological bent. France was under pressure in its possessions in Indochina, and the Dutch had lost the war in the East Indies. All of Africa was showing similar signs of anti-colonialism. It was a time of ferment.

We sighted Table Mountain in early April and were very soon docked, awaiting refueling and a fresh supply of groceries. We were only in port overnight, so I was unable to visit with my relatives, but did talk to them on the telephone. From the discussion with them and with others, it was obvious that apartheid lay heavy on the land. The National Party had supplanted the United Party in 1948, and the relaxed interracial attitude of those times had been supplanted by a strict adherence to the law and a forced separation of peoples of different race or of mixed race. Forced relocations of racial groups was the order of the day, and people were being designated as White, Coloured, Indian, and African, and thereby denied access across racial groups. There were stories of families separated from one another as a result of the decisions of special courts that employed the methods of the Nazis in measuring skull dimensions, hair texture, skin tone, and other factors to determine where people could reside. When I had visited South Africa in the war years and immediately afterwards, it had been a long way from a perfect paradise, but at least there had been hope. It was sad to contemplate the future. However, for visiting white men it was still a very pleasant place to be. Good food, excellent wine, lovely women, grand beaches—what more could we ask for?

We departed Capetown on a day that promised stormy weather as we proceeded eastward along the coast of Cape Province and Natal. It was quite a dusting that we received, courtesy of the admixture of the winds and the Agulhas Current. But as we got farther out into the Indian Ocean, the weather moderated somewhat, and we slowly passed south of Madagascar, then Reunion and Mauritius. Our route then took us south of the Chagos Archipelago, then on past the Cocos Islands and into Sunda Strait. There we passed the remains of the island Krakatoa which blew up in a devastating volcanic explosion back in the 1880s. The resulting tidal wave inundated the coasts of Java and Sumatra, causing heavy loss of life and property. (The devastating and deadly tsunami of December 26 2004, triggered by a shift in the tectonic plates that lie off Northern Sumatra and causing havoc around the Indian Ocean, had yet to happen.) We then entered the Java Sea passing Tanjung Priok, the port of Batavia, now known as Jakarta.

Our destination was Tanjong Gresik, the port of Surabaya, which lies east of Tanjung Priok on the north coast of Java. Tanjong Gresik was approached by entering a shallow water channel that lies between the islands of Java and Madura. We had the latest Admiralty charts on board, which were kept up to date with the latest information from notices to mariners. I had plotted the minefield channel on the latest edition of the nautical chart and had located the lighted buoys that marked the channel. We arrived off the channel entrance at about 3:00 A.M., and the Master was so advised. After awaiting his arrival on the bridge to no avail, I proceeded down the channel at slow speed. About five miles into the channel, we received frantic signals on an Aldis lamp from what appeared to be a small guard ship. Unable to interpret the signal, we nevertheless decided to retreat back up the mined channel. By this time, the captain had appeared on the bridge and we awaited the advent of dawn. Eventually, a pilot boat appeared and we proceeded down channel to the port of Tanjong Gresik. We never did discover what the guard boat signal meant, putting it all down to the confusion that still reigned in the former Dutch East Indies after the civil war that had raged there since the Japanese surrender.

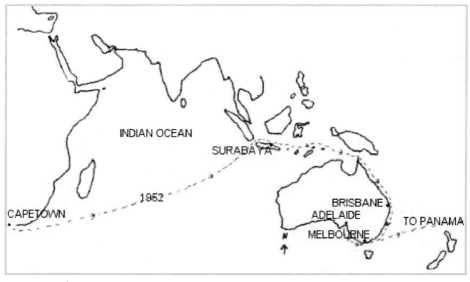

Figure 18 - *Angusglen* voyage around the Globe Part II, 1951/52

Tanjong Gresik was a busy little port that served East Java. The Dutch influence was still prevalent, with a Dutch foreman conducting operations at each of our cargo holds. But ashore, in the dock area, armed Indonesian guards were everywhere, roughing up the longshoremen when they detected broaches of our cargo of rice or for other perceived skulduggery. These people were not particularly well fed, and the stealing of a few pounds of rice for themselves and their families must have been a great temptation, even when these were just the sweepings from the hold. Our crew got very annoyed at this mistreatment of the longshoremen and, when well fortified with local hooch, carried out a spectacular raid on the local police station. Thank goodness no one was shot, their attack being so successful that the local constabulary fled the scene. Prisoners were released, and our heroes vanished into the night. When the police returned the following day they were still thoroughly demoralized and were unable to identify their attackers. It was a classic illustration of the civil and political chaos that was normal in Indonesia at that time.

I witnessed a rather clever attempt of robbery one day while inspecting our ship's hull from the earthen-filled pontoon that held the midship secured to the land. Our head and stern lines were fastened to bollards located on separate little islets covered in thick tropical vegetation. I noticed a small gang of men working around both bollards and discovered that they had cleverly cut off the eyes of three of the hawsers, and then secured them with twine to the one remaining effective hawser. In the event of a tropical storm, we would have broken loose of our moorings. As noted before, the local police were completely ineffective so, after chasing them off, we replaced the damaged hawsers and arranged for an armed guard for the rest of our stay.

For those who could afford it, there was ample food and lots of good Dutch beer to wash it down. There was a marine club in the dock area that served an excellent rijsttafel, a dish favoured by the Dutch in the tropics. In addition to the rice, it featured about a dozen other ingredients with peppers, curries, and other condiments. The additional bowls were each supported on the heads of rather attractive Javanese women who assisted one in selecting dishes. It was a grand way to eat in style while enjoying the local beauties. Washed down with several quarts of Amstel beer, it was a sure path to a lengthy afternoon siesta.

Upon completing discharge, we prepared our next voyage, which was to be to Australia to pick up mixed cargo for Eastern US and Canadian ports. We sailed, bound for Brisbane, Queensland, in late May 1952. Our route took us from the Madura Strait to the Flores Sea, passing north of the islands of Lombok, Sumbawa, and Flores. I can still see the dim lights of Malay fishing boats bobbing on the almost-flat, calm sea, and smell the perfume of spices from the islands. For that was where we were now—in close proximity of the fabulous Spice Islands.

Off Wetar island, close to Timor we ran into poor visibility at night, made worse by our radar deciding to become inoperative at a time when we were attempting to thread our way through many small islands into the Arafura Sea. All went well, and we carried on into the Torres Strait between Australia and New Guinea. At Cape York, the most northwesterly point of Australia, we picked up our pilot for the long run down inside the Great Barrier Reef to Brisbane, the capital city of the state of Queensland.

Shortly after leaving Surabaya our refrigeration system had gone on the blink, and the chief steward was forced to feed us all the best cuts of meat or lose them to putrefaction. During that passage to Brisbane, we lived high on the hog. Additionally, the chief steward engaged in a contest with several of the officers to produce a curry so hot that we curry lovers would cry uncle. It was a close-run affair at times, but when we approached the port of Brisbane it was still the officers winning the contest. We all, including the chief steward, looked forward to washing out our stomachs with gallons of good Australian beer. We were all very fit and raring to go ashore.

It had been a long haul from Cape York south to Brisbane, a distance in excess of sixteen hundred nautical miles. There were areas north of Cairns where the reefs made navigation difficult and we had to anchor until daylight. There was little to see, even when we passed the port cities of Townsville and Thuringowa. It had been hot at Cape York and around Cairns, but the weather became cooler as we went south, until it became a balmy seventy degrees Fahrenheit in the port of Brisbane. It was, of course, winter down under, and a welcome relief to all after our sojourn in the tropics.

We tied up at a wharf close to the entrance of the Brisbane River, in the port of Wynnum. The city of Brisbane itself was a short distance further up the river. It was very "British colonial" in layout and in the appearance of buildings, with a population of around four hundred thousand people. There was a good streetcar service between port and city, and pubs were not in short supply. The population was predominantly white, with a strong British connection. While preparing to load cargo, the first few days in port were spent mixing with the locals, drinking vast quantities of cold beer and feeling normal once more. Our cargo turned out to be bags of some special silicon sand that made good ballast for the voyages we would undertake around the east and southeast coasts of Australia.

It was early June when we departed Brisbane in excellent weather, heading south. The Gold Coast starts just south of Brisbane and was clearly visible on our starboard side. It was a small development in those days when compared to the massive stretch of properties visible today, but still quite impressive. Another missed opportunity to invest and make a fortune, like my 1948 visit to the Costa del Sol in Spain. Mind you, I still had not acquired that necessary initial capital.

By the time we were in the vicinity of Sydney, New South Wales, the weather had deteriorated considerably, and large seas, strong winds, and poor visibility were making navigation a miserable undertaking. Our radar was still inoperative, and the glimpse of large liners and freighters sliding close by in the gloom was not designed to keep one's stress at a comfort level. I called the Master to the bridge on several occasions, but he never appeared or replied, leaving me to sweat it out on my own. Strange behaviour from a ship's Master.

As we beat our way southward in rough weather, there now occurred an incident that could have ensured this epistle would never be written. I had a large bottle of cleaning fluid for the gyrocompass secured in my cabin by my bunk. Stupid place to have it, I know, but other bottles had vanished from the chartroom and the bridge, so I thought it would be safe in my own cabin. However, while I was sleeping, the cabin door came off its latch and slammed against the bottle, smashing it to smithereens, all brought about by the violent rolling of the ship in a heavy swell. The fumes from the cleaner fluid permeated the cabin and I went into a deep sleep, so deep that I missed both breakfast and my normal morning navigational chores. Someone must have refastened the door latch so the fumes were dissipated fairly rapidly. The only sign of damage was to the cabin deck where the paint had been lifted from the cement flooring. I was very lucky and knew it. There had been several reported cases of death from the fumes of spilled gyrocompass cleaning fluid, including one very bad situation on a Royal Navy cruiser, where several men had died in the gyro room.

We rounded the coast of southeastern Victoria State, and then passed through the Bass Strait in moderating weather. With King Island to port, we headed north westerly up Spencer Gulf to the Gulf of St. Vincent, where we entered the Port Adelaide. *Angusglen* then had to lie off at anchor for several days before a berth became available, this being in the days when the dock workers would go on strike at the drop of a hat. Eventually, we were accommodated and commenced loading our cargo, which consisted mainly of dried figs, dates, and other dried fruits. In addition, we took on a considerable number of cases of wine, which were stowed in the deep tanks for safekeeping. The dockers were friendly enough, but worked at their own pace, with frequent stops for cold beer and to place bets on the horse races going on all over Australia. We soon learned that "old bastard" was a term of affection, and "pommy" was recognition that you were regarded as an Englishman and acceptable, but that "pommy bastard" were fighting words. The beer, as usual in Australia, was good and chilled—it went down grand.

The city of Adelaide itself lay about ten miles or so south of Port Adelaide, but there was a good train service between the two places. The city was similar in size to Brisbane, with a roughly similar architecture—verandas and covered walkways everywhere to shade the sun. I went searching for good food but was disappointed in the offerings of even the better class of restaurant, and therefore decided to stick to ordinary grub, which was available in great quantities in pubs, cafés, and other plebeian establishments.

There was a nice little hotel fairly close to the port, in a place called Largs Bay Station. I had been told about the place by shipmates on *Salween* when she disembarked Australian troops in Adelaide during World War II. It lived up to its billing, and with the shoreline close by, was not unlike its namesake in faraway Scotland. I met a number of Aussies and was welcomed by them into their homes. The houses were of the same dimension as ours, with the same fittings, but without our double windows and insulation to keep out the cold. The toilet in one house, though, was unique, with almost a full library of books and magazines at one's disposal. I have never seen this idea duplicated anywhere else.

It was early July when we departed Adelaide for Melbourne, where we would complete loading. The cargo already in the 'tween decks was secured with wooden battens and wire fastenings to protect it from damage en route to Melbourne. We felt that at long last we were on our way home.

Chapter Nine

From Australia to Home via the Panama Canal

The passage from Adelaide to Melbourne was short in duration but rather eventful. A full gale developed shortly after we left Adelaide, which immediately exposed our feeble attempts to secure and protect cargo. The 'tween decks soon were in a state of shambles, and dried fruit floated in wine everywhere. Attempts were made to contain the damage, but the main effort would have to await our entry into a port of refuge.

Figure 19 - Entering Port Phillip Bay preparatory to loading cargo in the Port of Melbourne, July 1952

It was a great relief when *Angusglen* made it safely into Port Phillip Bay, and we were able to proceed quietly up the Yarra River to the port of Melbourne. It was now mid-winter in Australia, and quite cool and damp. The port lay about three or four miles west of the city of Melbourne, with excellent streetcar and train service between the two places. The mess in the 'tween decks was soon sorted out, and we commenced topping up our cargo. More dried fruit, but also a lot more wine, and a potential source of trouble—many hogsheads of rum and brandy. Special protected lockers had to be constructed to protect this valuable cargo, not just from the weather but from the possible depredations of both dockers and our thirsty crew.

Meanwhile, there was a chance to see both port and city, and I took full opportunity to explore both. There were many pubs in the dock area, with more squalor in the surrounding streets than I had previously seen in Australia. The city was different, with well laid-out streets and buildings, and many shops selling every conceivable type of goods. Additionally, there were good hotels and fancy eating establishments serving excellent food washed down with quite good local wine. A nice place to visit. I picked up some Australian ornaments made of mulga wood to take home to Doreen. As usual, I drank too much beer at a farewell party set up by some friendly Australians, and vowed never to drink again as I clung to the bridge railing as we sailed out of the Yarra River into Port Phillip Bay. It would be a long haul across the Pacific Ocean to the Panama Canal.

One incident that I witnessed in Melbourne still sticks in my mind after all these years. It took place in a dockside pub and involved Stan Hollis again, our worthy VC winner. He was in conversation with an Australian who turned out to be German and had been in the Western Desert with the 90th Light Division of the Wehrmacht facing Stan's 50th (Northumbrian) Division. In no time at all they were bosom buddies and holding one another up as they sang "Lili Marlene." What a difference to his attitude of wariness and suspicion with which he greeted the appearance of the Waffen SS blond supermen in that little river port in Argentina. He respected the ordinary German soldier both as an individual and as a fighting man. With the Waffen SS, he respected their fighting abilities—period!

Our course took us across the Tasman Sea to just north of the Three Kings Islands at the northern extremity of New Zealand's North Island. From there we headed east, southward of the islands of French Polynesia, to the vicinity of Pitcairn Island, from whence we steered a direct course to Balboa at the Pacific Ocean end of the Panama Canal. We were fully loaded and constantly butting into moderate to strong winds and seas, so our speed was reduced to a dreadfully slow eight and a half knots. It took us forty days to travel from Melbourne to the Panama Canal. A rather miserable voyage, as weather prevented maintenance work on deck, and the pounding into the seas never seemed to stop. The only highlights were the view of the infamous island of Pitcairn—of "Mutiny on the

Bounty" fame—and a bit of an upset when it was discovered that the crew had broached a couple of the hogsheads of rum. It was traced to the ship's carpenter who had access to the hold to conduct regular soundings of tanks and bilges, but who had used his opportunity to drill holes and siphon off considerable quantities of rum for sale to other crew members. Where there's a will, there's always a way. A tight guard was placed over the remaining hogsheads.

At long last, we entered the Panama Canal. Our refrigeration equipment had broken down again halfway across the Pacific Ocean, so it was decided that *Angusglen* would proceed through the canal to Colón, on the Atlantic Ocean, to carry out repairs and take on much needed fresh supplies. Our voyage through the locks and the crossing of Gatún Lake all took place without incident. The many varieties of colourful birds in flight showed up spectacularly against the green jungle backdrop and the smooth passage through each lock, under the control of motorized mules fastened to the ship, all helped to make it a worthwhile occasion. In addition, we saw a sign of home in the presence of a Royal Canadian Navy destroyer bound through the canal to the Pacific Ocean, en route to her base in Esquimalt, British Columbia.

We found Colón to be a fairly quiet and orderly place, as befitted its situation in the US-controlled Canal Zone. Everything appeared to be duty-free, with all brands of Scotch whisky going for very reasonable prices. However, Cristóbal, the Panamanian city next door, was a much livelier place with bars, clubs, and dives of all sorts open all day and night, with music and song filling the air with Latino presence. Naturally, I visited Cristóbal to soak up the atmosphere and the liquor, and to eye the female talent. However, I got a bit of a shock when I entered one dive and found that I was the only white person in the place. The occupants did not look the least friendly, so I skedaddled.

Within a few days, we were ready to depart from Colón to Baltimore, and headed northward across the Caribbean toward the Windward Passage that lay between Haiti and Cuba. We entered the Windward Passage through the Jamaica Channel, and then set a course from Navassa Island to take us clear of both Haiti and Cuba. From the Windward Passage, we set course for the vicinity of Bermuda, as our orders had changed and we were diverted from Baltimore to the ports of Quebec City and Montreal. Upon receipt of the change in destination, I advised my wife by cable to meet me in Montreal and to plan a holiday in New York.

It was early September when we approached Quebec City after our lengthy circumnavigation of the globe. It was pleasant to be home, but the ship's company wanted to complete our cargo discharge in a hurry so that we could terminate our long voyage in Montreal. The only person who wanted to stay on in Quebec was Greg Bellavance, our third mate, who was met by members of his family at the wharf. The poor fellow was ribbed unmercifully by his relatives, who commented upon his strong British accent, acquired through association for almost two years with a bunch of Scots and English.

Just after leaving Quebec, I was advised that Doreen was ill and would be unable to meet me in Montreal. After learning that she had to be operated on to remove a tumour from her bowel, I immediately arranged to leave *Angusglen* in Montreal and return to Port Arthur. I signed off ship's articles on the 8th September 1952, fully intending to return. But, unbeknownst to me at the time, I was never to see *Angusglen* again. Doreen was very ill for quite some time, and so another chapter of my life began.

Before commenting upon further personal adventures and strange happenings, it is now the time to briefly tell the story of the Canadian deep-sea fleet, which was vanishing very rapidly from the scene. During World War II, Canada built many deep-sea freighters to replace the horrendous losses of ships incurred at sea, particularly in the early years of the conflict. More than four hundred vessels of ten thousand five hundred dead weight tons plus forty coastal-type ships were constructed in Canadian shipyards during the war years. These were of British design and constructed by the Canadian government for bare-boat charter by the British Ministry of War Transport, or to be operated directly by Park Steamships, an arm of the Canadian government. After peace returned to the world in 1945, all of these ships were available for sale to bona fide Canadian companies. Most of these companies were established in Canada for the sole purpose of acquisition, and with little concern for continuity of ownership under the Canadian flag. Some of them were controlled by London Greek or New York Greek ship owners, who were motivated only by profit. Others were offshoots of British liner companies that normally traded into Canadian ports from Europe or Africa or the Orient. Then there was Canadian National Steamships, a Canadian government company allied with the Canadian National Railway. Finally, there were genuine Canadian companies prepared to continue in business under the Canadian flag, based on either coast.

Up until 1949, with a worldwide shortage of shipping and an unsatisfied demand for the transport of goods, most of these companies were able to keep operating at a profit, and indeed put money away for a rainy day. Even so, there was a gradual move to put vessels under other flags to save costs on crew wages and in the provision of other benefits so that, by late 1948, the deep-sea fleet was reduced to about one hundred and sixty vessels. These ships were all wartime-built, with fairly basic propulsion machinery. Now, they had to face strong competition from new, modern, faster craft coming off the stocks of the shipyards of Europe and elsewhere. Then, in early 1949, an ill-advised strike by members of the Canadian Seaman's Union enabled the companies, with Government of Canada support, to bring in another union to break the strike. It was a nasty business, with worldwide ramifications and, at the end of it all, the deep-sea fleet was reduced to fewer than sixty vessels. Now, in late 1952, we were down to about forty, so my chances of continuing my career under the Canadian flag were becoming slimmer by the month. A sad story of lack of political will to create a climate conducive to the retention of a small but viable Canadian merchant navy. In such conditions, even the bravest of Canadian ship owners succumbed to the pressure, and eventually the fleet was all gone.

Back in Port Arthur, Doreen gradually recovered and we bought a small house to ensure our privacy and comfort. To pay the mortgage on the house, I had to seek work locally and found myself back in the paper mill, hacking away at logs on the sorting table. To further augment our vanishing capital, I became a beer parlour waiter in the Mariaggi Hotel, a well-known Port Arthur watering hole, three nights each week. It was tough work some nights, fending off drunken loggers, but interesting to say the least.

Our house, though small, was not properly insulated and, in the very cold winter temperatures, required constant stoking of the wood and coal furnace. We both began to long for a warmer climate. When the paper mill job folded, I went back to the shipyard but did not enjoy the work outdoors in forty degrees below zero Fahrenheit weather. Luckily, through an old friend, Jimmy Graham, I was hired by the Canadian Car & Foundry Co. in Fort William to work as an estimator with their material management division, working on the building of training aircraft for the Royal Canadian Air Force and for North Atlantic Treaty Organization units. I knew nothing of estimating aircraft parts, job hours, and other essential elements, but I soon learned and, best of all, I was as warm as toast! It was now the spring of 1953.

Meanwhile, I attempted to keep up with what was going on in the deep-sea shipping world through the Canadian Merchant Service Guild. I was tentatively offered a job by the Guild to be navigator on a whaling expedition planned for the Antarctic, but this project failed to materialize, and I carried on estimating. Then, one day, I spotted an ad in the Guild's newsletter, where the Government of Canada advised that they were looking for ex-mariners with Foreign-Going Certificates who might be interested in becoming hydrographic surveyors. The only experience that I had relating to hydrographic surveying related to my time in cable ships, where we practiced precise navigation in both oceanic and coastal environments, but I felt that it was worth an application. Some weeks later, I was summoned to appear in Ottawa before a recruiting board. They even paid for my rail journey from the Lakehead.

The board was held under the jurisdiction of the Civil Service Commission, but was obviously influenced by the presence of a very large man who represented the Canadian Hydrographic Service (CHS), the organization that I would be joining if I could impress the board. The large man was Steve Titus, who turned out to be the Deputy Dominion Hydrographer. I was fairly comfortable with the questions forthcoming, but felt that I might have overdone my response when asked how I would feel about working in the Arctic. Thinking about my Murmansk experience brought forth my quite negative response to the question. I then compounded my error by waxing enthusiastic about Victoria, BC, where the CHS had a small regional office. I was given the distinct impression that Steve Titus was not impressed with my responses.

I returned to Port Arthur, unsure of my prospects and chiding myself for, as usual, being too outspoken. Then, in mid-June, I was advised that I was accepted as a technical employee of the Canadian Hydrographic Service and to proceed at my own expense to Victoria, BC. Obviously someone must have been listening to what I had to say. My life was about to change dramatically as I entered into a new profession. I departed Port Arthur by train for the West Coast, leaving Doreen behind to sell the house and to join me later.

Chapter Ten

Hydrographic Surveying the Waters of British Columbia

The lengthy rail journey from the Lakehead to the Pacific Coast was as always spectacular, and the passage through the Rockies and the Selkirk Mountains a grand experience. To build a railroad through Rogers Pass and other seemingly impassable barriers was a lasting tribute to the imagination and courage of the men who pushed it through, and to the many who died in doing it. The arrival of the train on the shores of Burrard Inlet was a fitting end to the journey.

A Canadian Pacific passenger vessel carried me from downtown Vancouver to Victoria Harbour. While en route, I absorbed the sights and sounds of the magnificent scenery of the Gulf Islands and Haro Strait before approaching Brotchie Ledge and entering the harbour. I reported immediately to the Canadian Hydrographic Service offices in Victoria. The district engineer was on sick leave and I found myself reporting to his deputy, Jim Brown, who was obviously a mariner and made me very welcome. He was an interesting man, having served in the Chinese Maritime Customs prior to the Japanese takeover and, later in World War II, conducting clandestine hydrographic surveys off the lower Burma and Malaya coasts in preparation for the planned allied landings in Johore Strait. The landings were never carried out, as Japan surrendered when the atom bombs were dropped on Nagasaki and Hiroshima.

I was introduced to members of the cartographic staff, but all the hydrographic surveyors were at sea. Arrangements were made for me to travel up the coast to Prince Rupert to join the survey vessel *Wm. J. Stewart*, which was operating off Porcher Island, just south of Prince Rupert. Meanwhile, I had one free evening in Victoria and meant to take full advantage of the opportunity. I found myself exploring Victoria's small Chinatown, and partook of an excellent meal at Don Mee's restaurant, an admirable establishment that is still going strong. While there, I met a relative of my brother-in-law in Port Arthur, and through him a small group of navy men and their ladies. I was then invited to a party, which continued at someone's house until the dawn was almost breaking. Great fun, but I had to get myself together to collect my gear and travel documents and get on my way to the north. This was all accomplished eventually—I can still see the knowing smile on Jim Brown's face—and I was on my way on the evening passenger ferry to Vancouver.

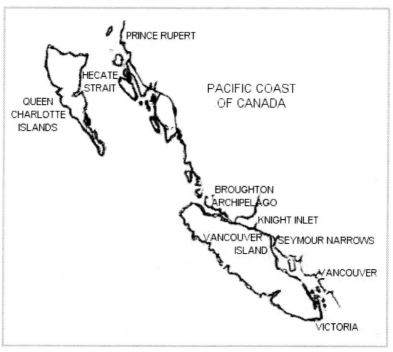

Figure 20 - The Pacific Coast of Canada

In Vancouver, I joined the Canadian Pacific passenger steamer *Princess Norah* and was soon underway up the Strait of Georgia to Seymour Narrows and beyond. The ship displaced about four thousand tons and was propelled by the usual triple-expansion engine. She cruised at about fourteen knots and was fitted with comfortable cabin space. I had a nice cabin and was fed exceedingly well in the dining saloon, with much assistance from an attentive staff of stewards. Like most vessels in the Canadian Pacific fleet, she had been constructed in Scotland especially for service on the BC coast. The passage up Johnstone Strait and Queen Charlotte Sound was pleasant, with lots of other marine traffic in view. These were mostly Canadian fishing vessels and US coastal ships bound to and from Alaskan ports. The run up from Milbanke Sound through the Grenville Channel was accomplished mostly in fog. Approaching Prince Rupert it had turned to rain, a rain similar to that in my hometown of Greenock where the locals, like those in Prince Rupert, all sprout webbed feet.

One ship's length ahead of our berth lay the CSS *Wm. J. Stewart*, my new seagoing home. She was painted white, both hull and superstructure, with a buff painted funnel amidships. Her length was 228 feet, she displaced 1,295 gross tons, and she was powered by a reciprocating steam engine that gave her a maximum speed of ten knots. She was built in Collingwood, Ontario, in 1932, and had a total ship's complement of sixty-eight persons, which included eight hydrographic surveyors, eight ship's officers and a crew of fifty-two. She had been fitted out with the latest hydrographic equipment in the thirties, but lacked any of the newer innovations in the field of electronics ushered in as the result of wartime advances in technology.

Figure 21 - CSS Wm. J. Stewart, 1953

I was able to get a steward off the *Princess Norah* to trundle my baggage along the wharf to the *Wm. J. Stewart's* gangway, where I was met by a young chap named Smith who advised me that he was an assistant hydrographic surveyor. He ensured that I met the hydrographer-in-charge, a bald-headed fellow in his forties whose name was Wilfred LaCroix. He assigned me to my cabin, which was located just forward of the surveyors' dining saloon on the deck below the surveyors' chartroom and promenade deck. It was not fancy, but pleasant enough with two large portholes to view the passing scene. Before sailing from Prince Rupert, I met the remainder of the hydrographic surveyors, and they were a varied lot indeed.

The senior hydrographer was also in his forties, with an abrupt manner that covered up his sense of fairness and decency. He was George Graves, a mariner who had served on Canadian Pacific passenger vessels prewar on both Pacific and Atlantic services. During WW II, he had served in the Royal Navy Reserve (RNR) and had been a gunnery expert providing advice to the planners of the

D-Day landings in Normandy. He had mustered out with the rank of commander, and it showed in his expectancy that his orders on board ship and away in the survey launches would be obeyed without question. He was a *Conway* boy, as was Jim Brown, the deputy head back in Victoria.

There were two other mariners in addition to myself. Len Kiernan was in his late thirties, a Liverpudlian to his fingertips. He had been with the Bibby Line, but served in the RNR during WW II, reaching the rank of lieutenant commander. He was a slaphappy sort of fellow, but jolly good company. He had married a local girl in Prince Rupert. The second mariner was Scottie Stewart, who had been with the Ben Line and the Union Steamship Company of New Zealand. He was a quiet man who seemed to have difficulty in expressing himself in the company of a group of self-assertive peers.

A most interesting man was Masao Saito, known as Mush. He was a civil engineer in his second year in hydrographic surveying, and very good at his job. He related well to his peers and the crew, being at ease with everyone. He was, however, of Japanese stock, although born in Canada, and deeply resented what had been done to his family and him during WW II. They had lost everything when they were moved away from the coast along with their neighbours, their property confiscated and sold for a small percentage of their real value to friends of the government. He had a justified case for feeling bitter, as he and his family were good Canadian citizens.

Additionally, there were two experienced hydrographic surveyors from Ottawa, on loan for the season to the Pacific Coast. Dusty DeGrasse was a qualified land surveyor from New Brunswick who had served on various East Coast–based ships. He was in his thirties and had been in the Royal Canadian Air Force during WW II. He was mostly a reticent type of man, but good company once you got to know him. The other import was Dick LeLievre, also a qualified land surveyor, but from Nova Scotia. He too was in his thirties and had served with the Royal Canadian Navy in WW II. He was a friendly chap and an extremely hard worker.

Next came an older man, Tom Hutton, a land surveyor by trade from Ontario. He was ex-RCAF from WWII. He was a pleasant fellow, but wondered why he was on board and obviously wished himself ashore someplace. He was joined by Dick Mooers, a civil engineer from New Brunswick who was in his late twenties but had little enthusiasm for the job. He also had a problem with alcohol that soon manifested itself.

Finally, there was Smithy, who had greeted me at the gangway, and another young chap called Woods. They were both students taking a summer job to earn their university tuition fees and to be exposed to some of the lore that pertains to nautical charting.

Over us all loomed Wilf LaCroix, the hydrographer-in-charge of the *Willie J*. He was an odd man in many ways, but he was dedicated to his work and meticulous in depicting our sounding efforts on the appropriate field sheet. He was not, however, a people-oriented man, and could rub his survey staff the wrong way without, I believe, fully realizing what he was doing. Luckily, he had in George Graves an excellent manager of men.

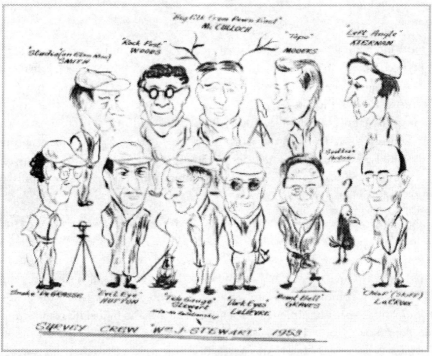

Figure 22 - The Clash of '53

Smithy drew a sketch of our motley group which could have been entitled "The Clash of 1953," rather than "The Class ...." In the sketch, I appear as "Big Elk from Back East" McCulloch, a reference to my celebrated persuasion of the manageress of the Elk's Club in Prince Rupert that we were all legitimate but thirsty members of the Elks and that I was a big Elk from Down East. Needless to say, this was at a time when all other drinking establishments had closed their doors for the night.

There were four 26-foot launches on board, three fitted for sounding operations and one for horizontal control and other tasks. The sounding launches carried Kelvin Hughes MS 14 echo sounders of the pre-war variety. They hardly represented the leading edge of technology, but were reliable and could easily be repaired by even the dumbest surveyor. The ship itself contained a Kelvin Hughes MS 14 echo sounder. Position fixing from the ship and launches was conducted by horizontal sextant angle taking of previously established points on shore. Hardly a high-tech operation, but quite capable of producing high-quality data for inclusion on nautical charts of the coast of British Columbia. The launches were capable of cruising at six knots and were seaworthy enough to operate in the most foul weather conditions to be found on the coast. They were crewed by a team of three—a coxswain, an engineman, and a seaman. Two hydrographers controlled the launch during sounding operations.

The actual day-to-day operation of the *Wm. J. Stewart* was the responsibility of the Master and his staff of qualified officers—two mates, chief engineer, three engineer officers, and a radio operator who also provided clerical support to the hydrographer-in-charge, who had the overall responsibility for financial and administrative matters affecting the ship. An odd arrangement, I thought then, but apparently it worked and had worked well for many years in Canada. Most other national hydrographic organizations operated with the hydrographers doubling up as ship's officers and with the captain in full command of both ship and survey operations. Our Master was George Billard, a good man and an excellent seaman. The mate was Ernie Betteridge, who took good care of the ship and the all-important launches. The chief engineer was a chap called McKenzie, who hailed from the Moray Firth. He had served with the Royal Canadian Naval Reserve in WW II and had a fund of stories to tell. He was efficient but had a drinking problem. The radio operator was John O'Malia, a bright fellow from Alberta, who had been in the Canadian Army during the fighting in northwestern Europe. He had a good sense of humour, even under trying conditions, and I looked upon him as a great shipmate.

Our working area was off Porcher Island, south of Prince Rupert, but fronting on the open waters of Hecate Strait. I soon discovered that it was launches away each morning, regardless of the weather conditions. Being bounced around in Hecate Strait was a daily ordeal that soon caused pain in both legs and knees in even the youngest and fittest of us. Combined with peering through the salt spray and the rain for a glimpse of a shore mark while wiping off the sextant telescope, it required considerable devotion and doggedness to survive each long day. This effort was followed by several hours on board working up our sounding and positioning data in relation to our tidal predictions. It was definitely not a job for the faint of heart.

Figure 23 - Our double-ended "covered wagons," 1953

Maximum spring tidal range was fast approaching so our operations were diverted into Kitkatla Inlet, accessed through Freedom Passage, which lay between the Porcher Peninsula and Goschen Island. The tidal rise and fall was a remarkable twenty-seven feet, so that entering the inlet at extreme low water presented one with a spectacular view of the uncovered kelp-laden rocky shoreline, towering way above our little launch as we navigated the remaining narrow channel of water that flowed out to sea at a fairly fast clip. After my time on the Great Lakes, where tidal effect is negligible, it was a truly astonishing sight. We mapped the extreme low water line using aerial photography provided, in the roughly two-hour time frame available. When we entered the inlet to conduct sounding operations later in the day, the entrance was broad and deep, with little hint of the hazards below.

I was rapidly learning on the job to conduct a variety of hydrographic surveying tasks and to process the data into a form that could be drawn onto a field sheet, which would represent the most accurate survey data for that point in time. The field sheet was very much the property of the hydrographer-in-charge, who guarded it jealously and only allowed access to it under strict supervision. The senior assistant was allowed some freedom of action, and he supervised our transfer of positioning data from the sounding boards to the field sheet. All inking of soundings onto the field sheet was a responsibility closely held by the hydrographer-in-charge. It created an atmosphere somewhat like a schoolroom, which caused great but suppressed hilarity among those of us newcomers who had been charged with considerable responsibility in our previous careers. I think that it was the attitude of the hydrographer-in-charge, who lacked rapport with his staff, which provoked our cynicism and determination to become truly professional in our new occupation.

The ship was a coal burner and carried enough fuel for six or seven weeks away from base but, by early July, we were getting low on bunkers, so we headed south for the vicinity of Johnstone Strait where we would be within easy striking distance of bunkering south of Seymour Narrows. In addition to conducting sounding operations in the strait, some time was spent up Knight Inlet laying in horizontal control and coast lining. It was pleasant to be away from the foul weather off Porcher Island; the inlet was peaceful, with deep water throughout and the mountains towering over the waters on either shore. Out in the strait, our launches were followed by pods of killer whales, no doubt curious as to our intentions. In mid-July, we steamed south to Seymour Narrows and passed the hazardous Ripple Rock with the strong tidal current in our favour. In no time at all we were past Campbell River and approaching Nanaimo, where we would bunker and conduct some revisory surveys in the harbour.

Our arrival in Nanaimo was planned for very early on a Saturday morning, so that all hands could have a break prior to coaling on the following Monday. The surveyors dispersed to be with their loved ones. I traveled to Vancouver to meet my wife, who was arriving by train from Port Arthur, before conveying her back to the Malaspina Hotel in Nanaimo. It was wonderful to be reunited, and the weather cooperated by being at its very best. On the Monday, while the *Willie J.* took on bunkers, we commenced our revisory survey of Nanaimo Harbour and Departure Bay. It was highlighted by a short drama that took place in the hydrographic chartroom on board the ship. One of the hydrographers was late in reporting for duty, being ensconced with his lady at that same Malaspina Hotel. It was Len Kiernan, that Merseyside firebrand. He duly apologized but was berated like an errant schoolboy by the hydrographer-in-charge. As one could have predicted, Lt. Commander Kiernan exploded and told him to ……off! We had lost a staff member and a good chap. I had my first doubts about my future in such an atmosphere. I loved the work—it was interesting and demanding—but such an overreaction from a superior was not a good harbinger of the future.

We completed the survey in Nanaimo Harbour and steamed south to Victoria to carry out some work off the Sidney Harbour wharf. I relocated my better half to the Sidney Hotel and so was reasonably content with life. At this time, I met Bob Young, the district engineer-in-charge of the Victoria office, for the first time. He seemed to be a nice enough fellow, a civil engineer to his fingertips, but perhaps lacking in the colour that one expects in those associated with the sea. Like LaCroix and others, he was a product of the Depression years, and as I gradually learned, it permeated his thinking. He was, however, a fair-minded man, meticulous to a fault in his work but, as I discovered later, with a highly developed sense of humour.

His responsibilities encompassed not just the Victoria office with its cartographic section, but the disposition of a small fleet of ships. In addition to the *Wm. J. Stewart*, the fleet had expanded to include the *Parry*, an eighty-seven-foot former coastal patrol vessel. She was a wooden-hulled vessel and was dedicated to tidal and current studies. Another addition was the *Ehkoli*, also a former patrol ship, dedicated to oceanographic studies. Finally, about to join the hydrographic fleet, was the *Marabell*, a former minesweeper and private yacht, which would conduct regular hydrographic surveys of the BC coast in support of the *Wm. J. Stewart*.

All of the foregoing was a vast change from the days of Henri Parizeau, who established the Victoria office of the Canadian Hydrographic Service back in the early 1920s. He had a fearsome reputation as a difficult man to deal with—at least Ottawa thought so, as he apparently managed the Victoria office as if Ottawa did not exist. However, most of his staff admired him, except perhaps any young woman in the office who attracted his roving eye. A man of many parts was Henri Parizeau! It was rumoured that the chief hydrographer of the day—William J. Stewart himself—was only too glad to have Parizeau out of Ottawa and in exile on the Pacific Coast.

I was slowly learning that the Canadian Hydrographic Service had an interesting history and perhaps was not as dull a place to be as I had originally surmised. Contact with DeGrasse and Lelievre, our surveyors from back east, gradually filled in the picture for me of the larger organization. I heard of *Acadia*, a predecessor of *Wm. J. Stewart* and numerous other hydrographic vessels that had been wartime minesweepers and frigates, and learned that Pictou, Nova Scotia, was the winter base for the East Coast fleet. At season's end—usually early November—the hydrographers packed away their accumulated data and returned with it to Ottawa, where the CHS headquarters were located. I looked forward to visiting Ottawa one fine day.

We departed the waters of southern BC in late July and made our way north past Ripple Rock in Seymour Narrows. I was becoming aware of the *Willie J.'s* connection to the infamous Ripple Rock. Apparently, in 1944, while stemming an adverse tidal current and attempting to avoid being swept onto the Rock, the vessel had nevertheless touched bottom and had to be hurriedly beached in Plumper Bay on Quadra Island farther north in Discovery Passage. The ship was badly damaged but salvageable, with much of her elegantly appointed surveyors quarters having to be replaced by more utilitarian woodwork. Only the hydrographic chartroom and the staff lounge retained evidence of the sumptuous décor of the thirties. Mind you, our below decks accommodation was quite comfortable, even if we christened the below-the-waterline cabin space "the Casbah"!

Figure 24 - The salvage of the *Wm. J. Stewart*, 1944

Our survey work continued without undue incident during the summer months, with most of our activity being concentrated off Porcher Island. Bunkering became essential in late August–early September, which happened to coincide with a national election. The hydrographers traveled by limousine from Union Bay to Victoria to do their civic duty and seek love and solace from their nearest and dearest. Doreen had settled into a commodious but older apartment and was making a home for both of us. She had found employment with the federal government and seemed to be quite happy, although missing her family.

We headed north again to the vicinity of Knight Inlet. It was a six-day week of work broken up by a quiet Sabbath in Telegraph Cove on Vancouver Island, where we watered and took on supplies and received mail. Actually, our routine was to arrive in Telegraph Cove early on Saturday evening, which just allowed us thirsty surveyors time to hire a water taxi to convey us to Alert Bay where beer could be had, together with women if you were so inclined, or a bout of fisticuffs even you were not. Here, one of our company managed to insult a young logger, which resulted in our man eventually sprouting two beautiful black eyes. He was known for ever after as "Dark Eyes LeLievre," and spent some time working in the hydrographic chartroom until he could see through a sextant telescope once more.

In late September the rains started, and soon the inlet was shrouded in a continuous downpour of cold wet rain, while streams of water cascaded down the steep slopes of the coastline. It was time to think about heading further south, but tradition dictated that our survey season did not end until mid-October, so we soldiered on, dripping rain over sounding boards while trying to see our horizontal control marks in the distance. Morale slumped and tempers became frayed as we struggled against the prevailing weather and wished the next few days would pass as quickly as possible. The hydrographers' lounge became our place of refuge, where bridge was played fiercely and without mercy. Then, at long last, we received the signal to proceed south to civilization.

The weather south of Seymour Narrows was quite decent so, of course, we had to stop off at Sidney to do a spot of work before docking in the Inner Harbour in Victoria. There was a cruel god somewhere making us pay for all our sins. However, on the 20th October, we at last tied up for our winter lay-up. It was good to be reunited with Doreen and, as Dusty DeGrasse and Dick Lelievre were departing shortly for Ottawa, she arranged an evening out for the two worthies and supplied them each with a lovely young lady companion from her office. It was grand fun, and those two shipmates of mine never forgot the occasion and often mentioned it when we met later in life.

Our place of work now became the Federal Building on Government Street, and we were each allocated offices where we could stow our equipment and the scads of data we had collected. My wife was working in the same building, so it was a rather convenient arrangement. We explored Victoria together while I settled into my office routine. After working up some of our survey data and meeting the cartographers, the chart correctors, and the tidal surveyors, there was not a lot to do. Nevertheless, I was advised to keep looking busy or my superiors would be most displeased. I discovered that there were no public drinking establishments in the city proper and that I had to go to Esquimalt, a neighbouring municipality, to get a glass of beer. Then John O'Malia, our ship's radio operator, advised me that just behind the Federal Building was a private club, known as "The British Public Schoolboys Club" or, to its members, as the Jokers, where a tipple could be obtained. I duly joined and became part of the scene.

Chapter Eleven

Enjoying the Local Scene – The Year of the Miracle Mile

Victoria was a grand place to spend the winter of 1953/54. The weather was mild, with no snow to speak of, but lots of rain sweeping in from the Pacific Ocean. The city itself was a jewel, with the legislative buildings and the Empress Hotel dominating the Inner Harbour. To take part in the Remembrance Day ceremonies and the parade afterwards led by the HMCS *Naden* band, as it left the Cenotaph and headed up the causeway toward the city, was to feel a part of the history of the country and the place. Much of the downtown area was tourist oriented, but many older, well-preserved buildings testified to the pride that people had in their history and their heritage. The Inner Harbour was still very much a working harbour in those days, with sawmills and shipbuilding and repair yards stretching up into the Gorge. Next door to James Bay was Beacon Hill Park, with beautiful gardens and ponds filled with ducks, and even a swan or two. The park fronted onto the sea, where it provided one with a superlative view of Juan de Fuca Strait and the Olympic Mountain range to the south, in the United States.

The city was also endowed with good live theatre, symphony, and ballet, together with a considerable interest in the other arts. It was the seat of the provincial government and in Esquimalt, to the west, lay the bulk of the Pacific Fleet of the Royal Canadian Navy and the Work Point Barracks of the Royal Canadian Regiment, which had just returned from service in Korea. Another shipyard plied its trade at the naval dockyard. Victoria was not an industrial city, but it did have a diverse selection of interests to sustain its vitality.

Hutton and Mooers departed the scene as anticipated, leaving a just a few of us to socialize from time to time. John and Helen O'Malia became firm friends, and we also spent time with Scottie and Smithy. We even travelled over to Vancouver by ferry to visit with my old shipmate, Len Kiernan, who was now into the land surveying business. It was a dreadful, wet weekend, with Vancouver at its very worst. On the way to church, my wife got her high heels stuck in tram lines and then had her high fashion hat blow off her head into a puddle as we climbed the steps outside the church. Altogether a day I have never been allowed to forget!

We made a few friends outside the hydrographic community—mainly through my wife's business relationships, but had yet to sink deeper roots. We loved the Saanich Peninsula, north of Victoria, where some of these people resided, with its rural atmosphere so like that of Cheshire in England, and started to think about a place of our own in such a setting.

Politics was beginning to interest me, particularly at the national level. It seemed to me that the Liberals had been in power for a long time and were losing their grip on the country. Under the leadership of Louis St. Laurent, they had won re-election in 1953, but they were exhibiting signs of arrogance and preoccupation, while retaining power at all costs. I fervently hoped for their defeat, but no one in the opposition Tory party seemed capable of persuading the electorate to sweep the rascals from power. At the provincial level, BC now had a Social Credit government led by W.A.C. Bennett, a long-time provincial Tory. I found it difficult to understand what Social Credit stood for, but I was even more confused by the conflicting claims of the various opposition parties. Obviously a matter for further study.

Internationally, there was good news and bad news. An armistice had at long last been concluded in Korea, but a peace agreement seemed to be a long way away. In Europe, there were stresses and strains between the West and the Soviets, illuminated by an East German uprising against the Communist regime. Oil was becoming an issue in the Middle East, as the Persians sought to take control of the Anglo-Iranian Oil Company and its resources. Meanwhile, Britain was continuing to gradually give independence to its remaining colonies, while life in large parts of the world appeared to be improving.

In the office, our time was largely wasted in silly repetitive tasks such as counting the number of charts in each box delivered from Ottawa. There was no training in techniques, not even in the surveying/cartographic interface. The lack of exposure to new ideas was truly a wasted opportunity for each of us, the Canadian Hydrographic Service, and the taxpayers of Canada. I simmered, but did not know how to bring about a change in attitude.

We were, however, starting to take on new staff in preparation for the coming season. Two seafarers, Ben Russell and Charlie MacIntosh, joined the *Marabell* contingent, which was under the command of Jim Brown. They were both worst. On the way to church, my wife got her high heels stuck in tram lines and then had her high fashion hat blow off her head into a puddle as we climbed the steps outside the church. Altogether a day I have never been allowed to forget!

We made a few friends outside the hydrographic community—mainly through my wife's business relationships, but had yet to sink deeper roots. We loved the Saanich Peninsula, north of Victoria, where some of these people resided, with its rural atmosphere so like that of Cheshire in England, and started to think about a place of our own in such a setting.

Politics was beginning to interest me, particularly at the national level. It seemed to me that the Liberals had been in power for a long time and were losing their grip on the country. Under the leadership of Louis St. Laurent, they had won re-election in 1953, but they were exhibiting signs of arrogance and preoccupation, while retaining power at all costs. I fervently hoped for their defeat, but no one in the opposition Tory party seemed capable of persuading the electorate to sweep the rascals from power. At the provincial level, BC now had a Social Credit government led by W.A.C. Bennett, a long-time provincial Tory. I found it difficult to understand what Social Credit stood for, but I was even more confused by the conflicting claims of the various opposition parties. Obviously a matter for further study.

Internationally, there was good news and bad news. An armistice had at long last been concluded in Korea, but a peace agreement seemed to be a long way away. In Europe, there were stresses and strains between the West and the Soviets, illuminated by an East German uprising against the Communist regime. Oil was becoming an issue in the Middle East, as the Persians sought to take control of the Anglo-Iranian Oil Company and its resources. Meanwhile, Britain was continuing to gradually give independence to its remaining colonies, while life in large parts of the world appeared to be improving.

In the office, our time was largely wasted in silly repetitive tasks such as counting the number of charts in each box delivered from Ottawa. There was no training in techniques, not even in the surveying/cartographic interface. The lack of exposure to new ideas was truly a wasted opportunity for each of us, the Canadian Hydrographic Service, and the taxpayers of Canada. I simmered, but did not know how to bring about a change in attitude.

We were, however, starting to take on new staff in preparation for the coming season. Two seafarers, Ben Russell and Charlie MacIntosh, joined the *Marabell* contingent, which was under the command of Jim Brown. They were both

Chapter Eleven

Enjoying the Local Scene – The Year of the Miracle Mile

Victoria was a grand place to spend the winter of 1953/54. The weather was mild, with no snow to speak of, but lots of rain sweeping in from the Pacific Ocean. The city itself was a jewel, with the legislative buildings and the Empress Hotel dominating the Inner Harbour. To take part in the Remembrance Day ceremonies and the parade afterwards led by the HMCS *Naden* band, as it left the Cenotaph and headed up the causeway toward the city, was to feel a part of the history of the country and the place. Much of the downtown area was tourist oriented, but many older, well-preserved buildings testified to the pride that people had in their history and their heritage. The Inner Harbour was still very much a working harbour in those days, with sawmills and shipbuilding and repair yards stretching up into the Gorge. Next door to James Bay was Beacon Hill Park, with beautiful gardens and ponds filled with ducks, and even a swan or two. The park fronted onto the sea, where it provided one with a superlative view of Juan de Fuca Strait and the Olympic Mountain range to the south, in the United States.

The city was also endowed with good live theatre, symphony, and ballet, together with a considerable interest in the other arts. It was the seat of the provincial government and in Esquimalt, to the west, lay the bulk of the Pacific Fleet of the Royal Canadian Navy and the Work Point Barracks of the Royal Canadian Regiment, which had just returned from service in Korea. Another shipyard plied its trade at the naval dockyard. Victoria was not an industrial city, but it did have a diverse selection of interests to sustain its vitality.

Hutton and Mooers departed the scene as anticipated, leaving a just a few of us to socialize from time to time. John and Helen O'Malia became firm friends, and we also spent time with Scottie and Smithy. We even travelled over to Vancouver by ferry to visit with my old shipmate, Len Kiernan, who was now into the land surveying business. It was a dreadful, wet weekend, with Vancouver at its very

Canadian Pacific Steamships men with wartime naval service in the RCNR. Additionally, an experienced ex-Royal Navy surveyor from the British Admiralty arrived, causing considerable speculation. Robert Sandilands had seen wartime service in the Fleet Air Arm. He was designated the Senior Hydrographer on *Marabell* reporting to Jim Brown. Then another mariner arrived. Ralph Wills had been with Canadian Pacific Steamships before the war and had served with the RNR during the war years before returning to deep sea and coastal trades at the end of hostilities. He was assigned to the *Willie J*.

All these men were a bit older than me, with the exception of Sandilands, whom I judged to be close to me in age. We now had a predominance of mariners in the CHS Pacific Coast Regional Office filling available hydrographer positions. Remarkably, four of them—Brown, Graves, Wills, and Russell, were products of the renowned training vessel *Conway*. Altogether, a wealth of maritime experience that I felt could do nothing but good for the Service.

A successful farewell dinner dance was celebrated, and on the 15th April 1954 we set off for Vancouver harbour and its surrounding waters, where we were tasked with conducting a revisory survey to bring the existing charts up to date to meet the cartographic compilation requirements for the publication of new editions.

Our area of survey included False Creek, Burrard Inlet, Port Moody, and Indian Arm. The surveys were conducted by launch inspection of the shore lines, making changes where necessary by eliminating out-of-date information and adding new information by resurveying approaches to new wharves and floats, and checking owners' plans and plots for chart accuracy. New large installations, such as wharves recently established on the north shore of Burrard Inlet, close to the First Narrows Bridge, and float complexes close to the Second Narrows Bridge, together with recent additions to the oil installations at Port Moody, took considerable time to survey. Most other areas presented little difficulty to the surveyor.

Our next survey grounds were to encompass the approaches to Sandspit and Skidegate Inlet from Hecate Strait, thence to the coastal area of Moresby Island, Cumshewa Inlet, and southward toward Cape St. James, a task that would take us more than one survey season. The weather was fresh but clear, and we found working in the Queen Charlottes to be mostly a pleasant affair, although it could be rough and unpleasant in the sounding launches, bouncing around off Moresby Island. There was an airport at Sandspit, which ensured that we received regular mail. Fresh fish was readily available over the side of a launch or dory, or from the nearby Haida villages. The Haida were largely a good-looking people, with attractive young women immediate attention getters. The men were strong, active characters mostly engaged in logging activities and in fishing using the well-maintained seine net fishing vessels that they owned. In past centuries, they had been a warlike tribe that attacked and pillaged the villages of other tribes to the south on Vancouver Island and beyond. The coming of the white man had brought disease and devastation to the islands, but the situation had improved in the last fifty years and a brighter future seemed assured.

Figure 25 - Launches ready for action, 1954

The fauna and flora of the Queen Charlottes were somewhat different from those on the mainland and on Vancouver Island. There were many plants and birds that were not exactly like their counterparts elsewhere but, not being a naturalist or a biologist and being kept fiercely busy with other tasks, I was never able to examine these differences at length. But we did view in awe the myriad sea birds that inhabited the many offshore rocks and islets, which also sprouted floral exhibits unique to the area. The tidal pools also displayed a variety of life more abundant than in the south. Some offshore rocks were the home of seal colonies and occasionally of a raucous, smelly bunch of sea lions. Life was never dull, whether watching red-beaked puffins drifting in the tidal current or noisy oyster catchers staggering around the beaches as they dug for oysters.

Launch parties often did not return to the ship at noon, but established a campground close to the sounding area on a suitable rocky outcrop or small islet. The doryman would be dropped off ahead of time to select a site and to prepare a suitable fire for a hungry launch crew. Food provided would usually include bacon and eggs, salami, bread, butter, and small cans of meat and fruit. Most of it would be fried or grilled on the blazing fire and eaten voraciously, even when sometimes overcooked or undercooked. On a wet day swept by the wind, the fire would smoke and embers fly as we huddled miserably together for shelter.

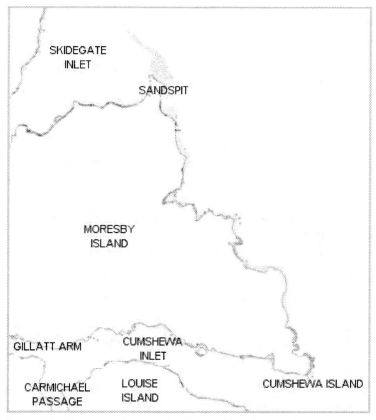

Figure 26 - Operational area in the Queen Charlottes, 1954

We once found a skull on the Skedans Islands offshore, but replaced it reverently as we realized that we must be on an old Haida burial ground. I also got a taste of the cold waters of Hecate Strait when my doryman managed to broach the dory while attempting to land my equipment and me on an unprotected beach. I felt like an idiot as I floundered in the waves, cursing the poor doryman and my fate in being involved in such an enterprise. Actually, I seemed to be permanently assigned to the launch *Curlew*, also known as *Port 2*, her place on the boat deck of the *Willie J.* The coxswain was a lad called Timmy, and the engineman was an older fellow by the name of George Becker, both of whom turned out to be completely reliable. The dorymen, however, came and went, with little training or respect for the sea.

On sounding operations, I was the right-angle taker and plotted and conned the launch along predetermined sounding lines while anxiously peering at the echo sounder trace for signs of a rapidly shelving bottom. My surveyor companion varied from time to time—some were quietly effective, some garrulous in the extreme, and some just plain obnoxious. I think that I was hard on them all, or so some have said in later life.

Our only break from the daily routine was the trip across Hecate Strait to Prince Rupert every second week for supplies. On the run across the strait and returning to the Charlottes, we would carry out a series of bottle station casts to obtain water samples and temperatures from various depths of water. These oceanographic operations were undertaken on behalf of Dr. Tully of the Nanaimo Oceanographic Laboratory, who was an old friend of Bob Young, our boss. Our trip across to Prince Rupert always appeared to be timed to get us into the port late on a Saturday night, when it was felt the ship's company would have little time to partake of the joys of Prince Rupert nightlife. Nevertheless, there was a mad dash ashore with yours truly in the van to get to the liquor store before closing time and to sample the offerings of the many hotels before they closed shop for the weekend. It reminded me very much my hometown, Greenock, where the pubs closed at 9:00 P.M. on Saturday and did not open again until Monday at noon. As in my hometown, too much liquor was consumed in too short a time, and the local constabulary were kept very busy. If the ship had docked at a respectable hour, the drinking would still have taken place, but in a much more relaxed and civilized manner. So sayeth the tippler!

Eventually, of course, we had to leave the Charlottes in preparation for our coaling operation. As usual, we spent a number of days operating around Kingcome Inlet and Minstrel Island in more protected waters. Minstrel Island was a particularly attractive spot, as it was on a regular seaplane route for mail and supplies and housed a small hotel that did a roaring business selling beer to thirsty loggers and seafarers. Once three of us—John O'Malia, Charlie Poole, and myself—sailed our little cutter many miles down Knight Inlet, all the way to Minstrel Island for an evening refreshment. It was a long drunken haul back up the inlet as the stars shone down upon us from a clear night sky. Needless to say, we had outstayed our welcome at the pub. We crept on board under the disapproving eyes of the quartermaster on watch. Poole was a new recruit to the hydrographic ranks, younger than the rest of us, and not really used to our rough-and-ready ways, but he soon learned.

Once again we traversed Seymour Narrows, and our trusty limousine awaited us at Union Bay to transport us home for the weekend. My wife was there to greet me and had some news that was not entirely unexpected, but nevertheless most gratifying for us both. She was pregnant with our first child. After eight years of marriage but apart for more than half that period, we were soon to be parents. It was a wonderful thing to contemplate, and we duly celebrated. Victoria was ablaze with flowers and shrubs, and the tourist season was beginning to make itself known in the Inner Harbour and nearby Beacon Hill Park. We sadly said farewell on the Monday afternoon, before I joined my shipmates for the long bus ride north, up Vancouver Island to Union Bay.

In our absence, the ship had been coaled, watered, and victualed, and was ready to sail on the first appropriate tide. In no time, we were through Seymour Narrows and up Johnstone Strait to our designated survey grounds around Kingcome Inlet. The weather was fine, and our work proceeded apace. A noticeable feature of most of the channels and inlets were solitary cabins hacked out of the surrounding forest or sitting on small floats at the water's edge. Each one was inhabited by a bearded and hairy man who lived in the wilds on his own, perhaps with the company of a straggly dog or wild-looking cat. Sometimes, their only means of transport was a small, flimsy rowboat they used for fishing and picking up supplies from Minstrel Island. They were truly loners and mostly did not welcome conversation. Each one, no doubt, had a tale to tell!

Then we were off to the Queen Charlottes again, this time to further survey Cumshewa Inlet and its nearest connecting channels. Our first task was to establish horizontal control using, where possible, provincial government topographic survey markers as a check against our own geodetic data. Sometimes, the provincial markers would be found in locations that only a seal or mountain goat could occupy. All this added to the fun of a day in the outback. Meanwhile, brown bears climbed the hills and stalked along the shoreline looking for fish and other varieties of seafood. Launch crews were supposed to alert the theodolite-bearing surveyor of the sighting and proximity of the animals, but sometimes their warning shouts and horn blowing came late in the game and only enabled a very perturbed surveyor to evacuate his station and make a stumbling retreat to the beach, with the bear in full sight and no doubt enjoying the scene.

Additionally, the hills were the home of a large number of small deer, which were also trying to avoid the predator. The hills themselves were a shambles of broken trees that represented the leavings after the hillsides were stripped of cedar for aircraft construction in WW II. It was a sobering sight, what we were doing to our planet. After ten years or so, little new growth was visible to the eye of the beholder. Surely we can do better in the future.

There was a logging camp at the head of Gillat Arm, a branch of Cumshewa Inlet, which attracted the bears like a honey pot. They could be seen hovering around the cookhouse and the garbage disposal area, seeking succour at most times of the day. At meal times, they would get quite bold and press themselves close to the cookhouse, only to be temporarily chased away by the cook and his helpers throwing missiles such as rocks, potatoes, and sticks at them and shrieking obscenities in various languages. It was a show well worth seeing, and gave us an opportunity to have live entertainment with our own cold plate lunch.

The approaching birth of our first child was prompting me to seriously consider my future. I liked hydrographic surveying, but promotion was liable to be slow unless I could obtain additional qualifications. My salary at present was insufficient to support a wife and family. I could seek further employment sailing deep sea, but with a smaller and smaller deep-sea Canadian fleet my chances of finding a suitable job were not too good. I therefore requested a leave of absence during the forthcoming winter lay up to enable me to study and then to sit for my Masters Foreign-Going Certificate. Permission was granted at a time of my choosing, but without pay or other support. At least I had a goal to aim at in early 1955.

The visit to Prince Rupert in August was a bit of a change of pace. In addition to picking up mail and supplies and emptying the shelves of the Prince Rupert liquor store, we conducted a detailed revisory survey of Prince Rupert and Port Edward and all adjacent waters. It was pleasant work, and lots of fun when we visited the fish plants with their large staffs of native women workers. The lasses, some of whom were rather attractive, were a jocular bunch and much given to sharp and pointed comments about the hydrographic surveyors invading their domain. It reminded me of similar experiences I had as a very callow youth in Scotland, when my duties as office boy and general dogsbody took me to the Gourock Rope Works. The women there were very rough-and-ready and opinionated and could frighten a young fellow half to death. The rest of our revisory survey was dull and technical and not nearly as much fun.

We thought back to our sojourn in Vancouver harbour as the Empire Games commenced in early August. The real highlight of these games came when the famous Miracle Mile was run. Two men, Bannister of England and Landy of New Zealand, ran the mile under four minutes—a world record for its day. It was a marvellous achievement and cheered us greatly as we enjoyed the attractions of Prince Rupert.

Meanwhile, the survey work continued in the Queen Charlottes and further south in and around Kingcome Inlet to the end of the survey season in mid-October. As usual, the rains dictated our location and the season termination date. The only untoward incident was a short period when Chief Engineer McKenzie fell off the wagon and went on a lengthy drinking spell that disrupted routine on board to some extent. He was a good man, a great engineer, but with this one great weakness. He had been with the *Willie J.* since 1947 and normally was a tremendous asset to her performance. In his cabin, he had given pride of place to a large wooden plaque of Adolph Hitler not, I hasten to add, in reverence to the man, but as McKenzie's booty when his Royal Canadian naval ship captured a German merchant ship off the west coast of South America. He eventually ran out of liquor, sobered up, and became a normal human being once more.

I was united with my very pregnant wife in mid-October, and we made plans for the arrival and my preparations for an attempt at obtaining that Masters FG Certificate. Andrew was born on the 9th January 1955, a lusty, bawling nine-pounder. His mother was weary but triumphant and his father was cock-a-hoop with pride, even if his son's first cradle was actually in the top drawer of a desk. Those were still the days of washable diapers, and soon myriads of drying diapers festooned our kitchen and other living quarters. My studies were sporadic and uncoordinated, and I began to fear for my upgrading prospects. My wife indicated that I should get that tuition I so badly needed, so I went off to Vancouver to join the School of Navigation. My routine became Monday to Friday in the school and home for the weekends. Not very fair to my lady, but that certificate meant future security. I found a little attic room in an old house in Kitsilano, where I existed on boiled eggs, bread, and milk. I was gradually becoming more confident of my chances as time went by, and my interaction with tutors and other aspiring candidates improved. In late March, I passed my written exams without trouble, but had to re-take the oral exam a week or so later. I was obviously a bit unconvincing the first time around, when Charley Barber, the examiner, trapped me off the coast of Mexico in a howling gale and I ran out of anchors to deploy to save the ship. The second time around, I was primed for his probing on my knowledge of tidal matters. I had read the book by Rachel Carson on "The Sea

Around Us" and had an inkling of the sort of answers he expected, and so it came about that I was awarded my Masters FG Certificate. If memory serves me correctly, I was advised by Stan Huggett, another mariner and now tidal surveyor, to read Rachel Carson's book. Anyhow, I was ready to return to the Canadian Hydrographic Service. It was now early April 1955, George Graves had departed to take a job in Ottawa with the Department of Transport, and Smithy had returned to school. Sandy Sandilands would be our Senior Assistant Hydrographer for the new season, and several new staff members were expected to join us shortly.

Chapter Twelve

Armed with My Master's Foreign-Going – Another Hydrographic Season

It was comforting to know that Doreen had made friends with several of the other surveyors' wives, who would be good company for her during my absence. She and Sue Wills seemed particularly close, I suppose with young children as a bond. Several new hydrographic recruits joined us on board the *Willie J.* prior to our departure from Victoria in mid-April. They were Paul Bender and Bob Durling, both Master Mariners, and Bill Robinson, a civil engineer and reserve fighter pilot. They were all good fellows, but with very different views on life and so, added to the stew of surveyors already on board, we groped our way toward some sort of cohesion and respect for one another. Thank goodness, most of them played bridge and, led by Sandy, we whiled away many an off-duty hour bidding and attempting grand slams with great abandon.

As usual, our first survey project was a continuation of the work around Kingcome Inlet, and then on to the Queen Charlottes and the coast, channels, and bays south of Cumshewa Inlet. Carmichael Passage ran south from Cumshewa Inlet, skirting the western edge of Louise Island, before opening up into Lagoon Inlet and Sewell and Selwyn Inlets. It was all new territory to us, and without obvious signs of devastation. Inevitably, our launches found shoals by error of judgement and rescue operations had to be undertaken. From Selwyn Inlet we spread out into Dana Passage and Dana Inlet and then out into the more open waters of Laskeek Bay, Reef Island, and the Lost Islands. During a coastline survey, my doryman managed to strand our dory several hundred yards inland on a long stretch of beach out on Titul Island, much to my displeasure. Then, in Logan Inlet, while operating my theodolite on Flower Pot Island, I managed to fall off the survey station over a steep rock and soil slope, together with instrument, down to the narrow beach. It was a long fall, probably at least seventy feet, and it knocked the breath out of me, and bruised my body from shoulder to thigh. However, no bones were broken, the theodolite survived, and only ego was affected. Not a soul in the launch spotted my debacle and came to my rescue. All in all, a most undignified episode.

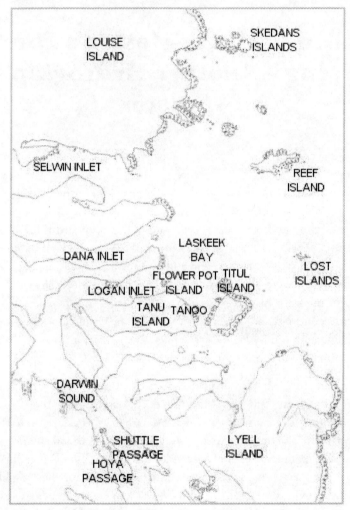

Figure 27 - Operational area, Queen Charlottes, 1955

Another ex-mariner joined us at Sandspit. His name was Al Ages, a Dutchman from Friesland who had served at sea mainly on the passenger ships running between Holland and the East Indies. He was a likeable chap, with a highly developed sense of humour. He fitted in well with our little group and enlivened the Casbah, the lower deck part of the surveyors' quarters. Later in the book, I will describe our association in the ice-covered waters of the Western Arctic.

Figure 28 - Poole, Sandilands, Wills, with Al Ages in front, Butedale 1955

Just to the south of Laskeek Bay lay Tanu Island, which was the home of the ancient Haida village of Tanoo. The village had been abandoned around the turn of the century, after the population was ravaged with the effects of smallpox. The wooden buildings and totem poles stood as a silent memorial to these brave people. Awe inspiring and thought provoking, ushering in contemplation of the ghosts of the natives looking on in silent reproach.

At about that time I finally got a sensible doryman who took an interest in what I was doing and was not afraid to ask questions. His name was Willie Rapatz, and he was a young man from Austria. His English, while accented, was clear and understandable, the result of schooling and some time working with the British Army of Occupation in Austria as a translator. It was a pleasure to have someone to converse with while carrying out my survey duties. He was rapidly promoted to bowman and then coxswain, and after a few years joined the hydrographic staff, eventually going on to head the Pacific Coast Regional Tides and Water Levels Division. Today, we are both members of a small but exclusive group of former hydrographers who meet monthly to exchange views and insults while deploring the present state of hydrography in Canada. Who could have guessed such an outcome from our first casual meeting on some lonely beach in the Queen Charlotte Islands?

Figure 29 - Crew of *Port 2*, Queen Charlotte Islands 1955

Our survey work varied between exploring the many channels and inlets between islands and carrying on with the hard slog of sounding operations offshore with the twenty-six-foot double-ended canvas-covered launches twisting and rolling in the lively sea. Good eyesight and good legs were all important while seeking a glimpse of a distant whitewashed rock or a flag on an often-obscure shoreline. The air was blue with profanity as sextant mirrors misted up with salt sea spray and coxswains were berated for not holding to a pre-selected traverse. Wearying but exhilarating, as each sounding launch would vie with the others to claim the highest sounding mileage for the day's operation.

One coxswain I had on *Curlew* was named Mel. He was a large, even-tempered man who wore very thick glasses and really could not see very well. Getting him lined up on a distant shoreline and picking out a proper transit that he might follow was a frustrating experience as our plotted track weaved about like a drunken sailor. Luckily, Mel soon tired of being coxswain of a launch and left the *Willie J.* for other pursuits.

The surveys in southern waters went without a hitch and in mid-October we were heading south once more through Seymour Narrows toward our winter lay-up in Victoria. Scotty left us to go to New Zealand, and Bill Robinson joined a civil engineering firm. John O'Malia left to go to work as an electronic technician at the Naval Dockyard. Otherwise, our group remained stable, but bored in the confines of the Federal Building. It was grand to be home with family and friends, but the lack of a proper plan for long-term training was a crying shame. It was true that those of us judged to have a decent fist for inking in soundings could involve ourselves in the more delicate arts of chart compilation, but nothing was being done to prepare us for the coming explosion in technology and technique. I could not afford to take further time off to seek additional qualifications, such as a university degree. With a youngster at home and my wife devoted to his upbringing, my pay cheque barely met our minimum needs. I sought other sources of employment, but deep-sea jobs were drying up and coastal traffic was also diminishing. I put out feelers regarding the weather ships and the naval auxiliary vessels operating from the naval dockyard, but vague promises were all I received. I even toyed with the idea of getting into the cargo management business, where I knew several ex-mariners had done fairly well, but could not make the right connections. It was a slightly concerned person who prepared for another survey season in 1956. A small promotion helped a bit financially, but the future looked rather uncertain.

Chapter Thirteen

More of the Same – But Changes in the Offing for 1957

As usual the *Willie J.* commenced her season operating around Johnstone Strait, Tribune Channel, Minstrel Island, and Sullivan Bay. Our new radio operator/electronic technician Joe Haegart was making himself known as a very independent-minded individual. He had been with the Canadian Pacific on the Asia run prewar and had lots of tales to tell of the goings-on of the celebrated passengers who took passage on his ship. The names of people like Somerset Maugham, the author, and Gloria Swanson, the actress, came in for a fair amount of critical comment. Joe was a peculiar mix of fundamental Christian and Prussian military genes, an explosive spark that sometimes ignited in spectacular fashion. His confrontations with Wilf LaCroix, our hydrographer-in-charge, were a sight to behold. Life could never be dull with Joe on board.

John Bath, another ex-mariner, joined us from the Tidal Division. He was a rather quiet man of few words, with interests beyond the day-to-day sounding operations. An unexplained anomaly on the sounding roll could claim his full attention, even as we hurtled toward shore at the end of a sounding line. That could be most disconcerting! Later in the year, in the Charlottes, he caused our launch to go up on the beach as he countermanded my order to the coxswain to go hard a-starboard by yelling "hard a-port." The coxswain was so startled that he drove us up on the beach. I obviously needed a more reliable left angler.

Confrontations or differences of opinion between surveyors in sounding launches were not uncommon, probably as a result of our maritime backgrounds. A famous example was the occasion when Russell and Bender tangled over control of a launch and wrestled one another for possession of the three-arm plotting protractor. In the ensuing struggle, the plastic instrument flew out of their hands and finished up in a watery grave. Childish behaviour of course, but with strong-willed individuals under stress every day, not all that unusual.

In the Queen Charlotte Islands, we continued our exploration of the channels and inlets south of Louise Island, and sounding offshore out into Hecate Strait. We were now surveying down Darwin Sound, into Shuttle Passage, Hoya Passage, and around Lyell Island in the vicinity of Bigsby Inlet and Beresford Inlet. In the middle of all this activity, Ralph Wills departed on a special assignment to the Western Arctic related to the construction of the Distant Early Warning radar system. He was to liaise with the US authorities conducting hydrographic surveys on the coastal approaches to selected sites. A replacement for Wills arrived, a chap called Peter Ainsworth, ex-mariner with the Canadian Pacific. He was tall and gangly, and sprouted a full head of red hair, which was a clear giveaway to his volatile personality. He also had a commercial pilot's license, which stood him in good stead later in life.

On the east coast of Canada, our new ship *Baffin* had been fitted with a Decca 6f electronic positioning system, and we were advised that the *Willie J.* would be next on the list. Therefore, plans were drawn up to establish a temporary Decca Chain to cover Hecate Strait. Two Decca stations, master and slave, would have to be established by horizontal control in the 1956 season and the sites cleared as far as possible prior to the actual siting and monumenting of the electrical centres, and operation of the chain in 1957. The sites selected were on East Copper Island in the Queen Charlotte Islands and on McKenny Island in Caamano Sound on the mainland shore of British Columbia.

East Copper Island lay on the seaward side of the Copper Islands, located in Skincuttle Inlet, south of Juan Perez Sound and Burnaby Island. It was difficult of access by boat and thickly wooded. These trees masked an extremely rugged and difficult terrain that inhibited the creation of the suitable electronic mat required by Decca. It required a major effort by the ship's officers and crew to ensure success.

On the mainland coast, we established ourselves off Aristazabal Island to the south of Caamano Sound. A group of islets south of Moore Islands provided the *Willie J.* with a safe anchorage while we surveyed the area and assessed McKenny Island's possibilities as a Decca site. McKenny was not heavily wooded and was fairly flat on top, so it easily met our required criteria. Horizontal control had to be extended from Aristazabal Island and other sources, so an expedition was mounted to climb Mount Parizeau to recover a control point established by CHS back in the twenties. It was a major hike inland from the beach, hacking our way through thick salal and small trees while attempting to skirt small lakes that did not show up very well on the aerial photograph provided. Beyond the lakes, the land cleared somewhat as it rose imperceptibly before climbing fairly abruptly to culminate in the peak of Mount Parizeau. It had been an exhausting climb for our team of Sandy, Chivas, and me.

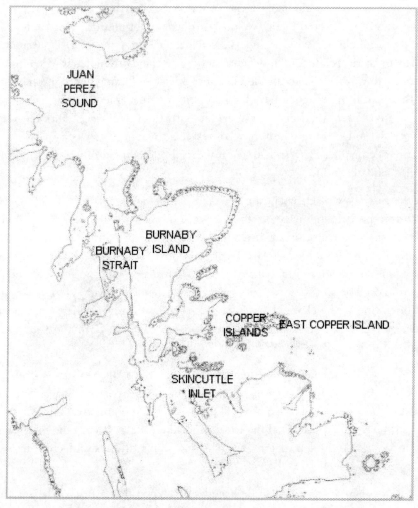

Figure 30 - Operational area, Queen Charlottes 1956/57

After a brief rest, we searched the area for a hydrographic benchmark, but only found the remains of a small wooden tower. We carried out a few observations before descending from the peak. By this time we were bone dry having consumed all our water. Chivas succumbed to thirsty temptation and fell into a muddy, stagnant, swampy pool. The rest of us pressed on, fearing all sorts of horrible bugs. It was about 10:00 P.M. when we staggered out of the forest onto the beach and were picked up by the launches from the ship. We had left on our expedition before 8:00 A.M. that morning, so it had been a very long, tiring day with little to show for our pains.

Eventually we managed to tie McKenny in properly to the horizontal control network and were able to prepare the island for the setting up of a Decca station. We then commenced a hydrographic survey of the west coast of Aristazabal Island and the islets and rocks in the vicinity of Moore Islands It was a messy area to survey, with lots of submerged rocks and reefs, not all marked by telltale streamers of kelp. The tidal rise and fall, although not quite as large a range as at Porcher Island, was still close to twenty-four feet at springs. Low watering was interesting but could be hazardous. I left Bob Durling on a rock that would be awash at extreme high tide and instructed him to carry out observations while I completed surveying an adjacent low water area. When I finished my task and returned to Bob, his rock stood only six feet clear of the water and he was ready to swim. It took me some time to live that one down.

A pleasant reward on these low watering expeditions was the search for abalone, which could be found a foot or so below the water surface in rocky areas were exposed to the elements. A few abalones with muscle broken were delicious when fried in butter. The anchorage where the *Willie J.* lay was a superb fishing ground, and many large halibut were hooked and drawn to the surface. A common sight of an evening was to witness a dory crew wrestling with a large halibut on the surface and watching in helpless laughter as the halibut landed in the dory and its occupants fell into the sea. Oh, happy days!

The rest of the season proceeded as expected, without undue incident. Our bridge games continued each weekend away from port. Weekdays, we were too exhausted by the time data had been processed after supper to do anything but collapse on our bunks. Sandy was a keen player and his Edinburgh accent came to the fore as he led the bidding. He was our most experienced hydrographic surveyor and his years with the British Admiralty were an asset of considerable value to the survey. He was also amiable enough in his dealings with the hydrographer-in-charge to shield the rest of us from possible confrontations or misunderstandings. He was a keen pipe smoker and could fug up the lounge in short order. He also had a fetish regarding poached eggs, which were often delivered in front of him at breakfast on a plate that contained the eggs swimming in greasy-looking water. As his seat mate, I always braced myself for his inevitable explosion. I hope his good lady June has long since mastered his exacting requirements for poached eggs. Back to bridge again for a moment—as an illustration of how weary the long hours of surveying could make one —I once answered a bid by shouting "FIX!"—I had been operating a sounding launch for far too many weeks!

We were all thankful in mid-October when we headed south again for winter lay-up. It was good to be with family once again in the little duplex that we rented close to the Jubilee Hospital. I had acquired a small car—an Austin A40—and, after some help from Sandy and others, I was licensed to drive in British Columbia. We were mobile at last. The winter went quickly, with a fair amount of socializing. In the office, I had progressed to being trusted to work on a fair sheet, the step before compilation in the production of a nautical chart. It required painstaking attention to detail, selectivity in the retention or elimination of data on the rough field, and finally a good fist, or drawing ability. A bit boring after a while, but at least I was not standing around trying to look busy. A pleasant interlude during the winter was the arrival in town of my brother Gordon for a short visit. He was on his way home from Malaysia to Scotland. He looked fit and was no doubt enjoying his life as manager of a rubber plantation up the bandit country of Perak, where a Communist uprising was making life difficult for all concerned. We got together with old friends from Greenock and sank a few.

We sailed as usual from the Inner Harbour in mid-April 1957, loaded down with supplies and material for the Decca stations and the technical staff who would set up the stations and operate the equipment. It had been a particularly wrenching parting from my wife, as she had informed me that another child was on the way. I had a spur to work to ensure that the project was successful and that my efforts would hopefully be recognized.

We headed straight out into Queen Charlotte Strait after clearing Johnstone Strait, on a course that would take us directly to Cape St. James and our first destination—the Copper Islands. It was a wild voyage, as we ran into a howling gale that caused minor chaos on board. The ship had spent so much time in reasonably calm waters that many items on board were not fastened securely. Below decks the alleyways were awash with broken cartons and boxes of food supplies intermingled with wet flour, broken raw eggs, and vegetable oil. It was an awful mess and took some time to clean up when we eventually reached a safe anchorage off the Copper Islands.

Our major task was to complete the logging of the Decca station area and establish the geodetic position of the station's electrical centre. A look at the accompanying photograph will give the reader some appreciation of the difficulty of the task.

Figure 31 - Decca 6f trials and tribulations, 1957

However, the job was done and we steamed across Hecate Strait to McKenny Island in Caamano Sound to complete the chain. Of course, while the ship's company were engaged in the construction of the Copper Island Decca station, our hydrographic launches were busy continuing our work of surveying the channels and bays and the outer coast line in the vicinity.

Figure 32 - Our young eagle "Joey," 1957

There were many eagles with nests on the larger trees and rocky promontories on the coastlines of the many islands. I recall sounding in one channel while a big eagle dove over the launch and seized a huge salmon in its claws before flying off with its prize to a nearby nest. Startling and awe inspiring! One of our launches found a young eagle that appeared to have been abandoned by his parents. He was christened Joey and became quite a fixture on board the *Willie J*.

By late May, our Decca chain was properly calibrated and we were able to commence sounding operations by ship, with the ultimate goal of providing sounding coverage of all Hecate Strait. The usual breaks for coal-ups took place roughly every six weeks, which gave us all a short spell at home during the summer. Our daughter Ellen was born on the 30th August 1957, so my responsibilities mounted. Even so, I was silly enough to get into a confrontation with the hydrographer-in-charge that brought me close to resigning. Luckily, tempers cooled and I determined to keep my feelings under control until I had decided on my future.

Sounding operations using Decca for positioning were usually dull affairs as we plodded back and forth across Hecate Strait, running our sounding lines east and west, starting up toward Dixon Entrance and working our way south to off Sandspit and the southern half of the strait. It was dull work and boring, particularly as the sea bottom was largely flat and uniform. We watched, however, for possible lane fluctuation, bearing in mind the awful disaster that afflicted the CSS *Baffin* on the east coast that same summer of 1957. She had relied too much on her Decca chain readings and managed to go up on Black Rock while it was in clear view.

Life was not entirely grim and miserable on board the good ship *Willie J*. I do recall my introduction to Bols, the Dutch gin. It was after the conclusion of survey operations on a Saturday afternoon and we were bound for Butedale across Queen Charlotte Sound in increasingly rough weather. Al Ages invited me down to his cabin in the Casbah and opened up his expensive Bols as a treat for me. I immediately let myself down in his eyes by asking for a mix such as orange juice. While imbibing and yarning and downing the Bols, I became ever more conscious of the movement of the vessel and the overpowering smell of an improperly cured caribou hide that took pride of place on his bulkhead. Blame the ship's movement, the Bols, or the caribou hide—I soon departed to the soothing comfort of my own bunk.

Then there was the occasion on the last day of the field season when my launch crew loaded up the launch with a number of cases of beer picked up at the Minstrel Island pub. They intended to have a good celebration en route to our home port. All was well until we went alongside to be lifted on board. Our extra weight was soon detected and the smuggled beer was confiscated amid much wailing and recrimination. Captain Billard was not a heavy-handed skipper, but he had no choice in the matter. I must admit that I liked George Billard as a man and as captain. Many a watch I shared with him when the *Willie J.* was underway, just chatting and exchanging views on a variety of subjects. I was saddened at his death many years later.

Once again we were happy to tie up in the Inner Harbour of Victoria for winter lay-up. Ralph Wills had returned from another spell in the Arctic with the US hydrographic surveyors, carrying out surveys of the approaches to the beaches nearby DEW Line sites. He finished his project on the USCG vessel *Storis* when she and two other coastguard cutters, *Bramble* and *Spar*, rendezvoused with HMCS *Labrador* in Bellot Strait before going on to circumnavigate the Arctic. Mike Bolton from CHS headquarters headed up a team of hydrographers on *Labrador*. It sounded very exciting to me, and I wished that I had been there.

A number of changes occurred in staff positions during that winter lay-up of 1957–58. Jim Brown came ashore to look after Sailing Directions and Sandy went to take over as hydrographer-in-charge of *Marabell*. Ralph Wills became senior assistant on the *Willy J.* and I found myself preparing to field a small launch party to survey Ganges Harbour on Saltspring Island, to be followed by an assignment with the US Coast Guard in the Canadian Western Arctic. It was all invigorating, and the promise of promotion made it all worthwhile.

Chapter Fourteen

New Responsibilities – Ganges Harbour and the Western Arctic

It was fun to prepare for the survey of Ganges Harbour on Saltspring Island. I had been assigned an assistant who was quite a bit younger than me. His name was Marvin Oro; he was a recent graduate of the Southern Alberta Institute of Technology and was a great companion and worker who contributed much to the success of our venture. Marvin was no seafarer, but he had been well trained in surveying theory and techniques, which he deployed with effect. We quickly gathered our equipment and survey documents together and were assigned a rather ancient twenty-six-foot-long wooden-hulled survey launch. She was a double-ender with a canvas covering similar to those I had become familiar with on the *Willie J*. A crew of three were recruited, none of whom appeared to have much experience. With no bands playing and few spectators, we slunk out of Victoria harbour and sailed uneventfully up the coast past Sidney before entering Ganges harbour.

Our base of operations was the Harbour House, a rather rustic structure located at the head of the harbour and close to the government wharf where we tied up our launch. The hotel was owned by the Crofton family, who made us very welcome. The family were one of the first settlers on the island and known to all. Desmond Crofton had been commanding officer of the Canadian Scottish regiment in WW II, and still conveyed a take-charge attitude in his day-to-day affairs. The accommodation provided was not luxurious but comfortable, and we had a small plotting room provided in the premises. The only drawback was that our rooms were located immediately over the beer parlour, which could become quite noisy later in the evenings. Being weak souls, we sometimes gave up working of an evening and joined the convivial company filling the beer parlour. The food was quite good, if I recall correctly.

Ganges harbour and approaches had been surveyed by the British Admiralty in the 1880s and was named after their ship HMS *Ganges*, an explanation of the harbour's naming after the enormous river that flows through much of Northern India. There had been many changes to the port of Ganges over the years, but most of the harbour was little different from the way it had been back in the last century. We set out to establish a horizontal control network and an automatic tide gauge and very shortly were ready to commence sounding operations. While the foregoing work proceeded, I wrestled with the cumbersome purchase system that I had to use to obtain lumber, lime, nails, gasoline, batteries, and spare parts for the launch. Everything had to be done by purchase order or by depleting my tiny allocation of petty cash. Very frustrating, particularly as the local merchants were reluctant to await payment by the slow processing of purchase orders, and the launch engine constantly needed such items as spark plugs and other vital parts to function. Nevertheless we accomplished much, surveying and plotting the many wharves and floats and sounding the outer reaches of the harbour.

There was an opportunity to explore Saltspring Island itself, which was a very attractive place with a mix of high lands and lower meadows, with of course the usual dominating coverage of cedar and fir trees. The population at that time was about two thousand souls, mainly concentrated in and around the settlements of Ganges, Vesuvius, and Fulford Harbour. Most residents were retired people from other parts, together with some farmers, fishermen, and local businesses. It was a pleasant place to be, and within easy striking distance of Victoria once one was off the island. A good ferry service was in place between Fulford and Swartz Bay on the Saanich Peninsula, Ganges and the other Gulf Islands, and Vesuvius and Crofton in the Cowichan Valley. My wife visited us one weekend, escaping our children, which added to the pleasure of our stay in Ganges.

In late May, we were recalled to Victoria to prepare for our Arctic venture. The launch was sailed back to our depot and the need to have a thorough engine overhaul was emphasized to the depot staff. We then found that we were going to be exposed to a new piece of electronic equipment prior to our deploying it in the Arctic. This piece of equipment was the tellurometer, manufactured in South Africa, and now to be made available for our benefit. The Government of British Columbia, Surveying and Mapping Division, had received similar equipment the previous year, and were prepared to train Oro and me in the intricacies of these distance measuring instruments and in their deployment under harsh conditions in the wild. They were fitted with excellent radios, which made communication between stations relatively easy. Eventually, we received orders to proceed to Seattle, where we were to join the US Coastguard Icebreaker Cutter *Storis*, which would be leaving shortly for Kodiak, Alaska, and then on into the Canadian Western Arctic.

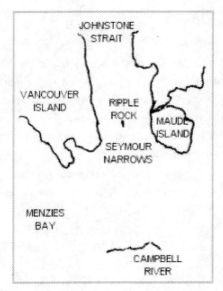

Figure 33 - Ripple Rock in Seymour Narrows

While we prepared for our Arctic project, an event of some significance occurred. Ripple Rock, the hazard lying in the middle of Seymour Narrows and the nemesis of the *Wm.J. Stewart* in 1944, was blasted out of existence in the largest man-made non-nuclear explosion to that time.

We would love to have been directly involved in the hydrographic survey before and after the event, but watched the whole affair on television in the comfort of home in Victoria.

Figure 34 - Explosion at Ripple Rock in 1958

In late June, the weather in Seattle was stifling with little wind. After an abortive trip to the shipyard at Bainbridge Island, we finally located *Storis* at Navy Pier 90 back in Seattle and were welcomed on board. Most officers and crew were still away on leave, so the welcome was relatively subdued. Each of us was to share a cabin with a junior lieutenant and, inevitably, found himself in the top bunk. There was no place to stow our equipment, and working space was supposed to be in the guardhouse by the gangway, which was presently occupied by an armed guard. We then set off to find our tellurometers, which were apparently in the hands of US Customs. These exotic instruments were being held, as they were manufactured in South Africa, a country infamous for its apartheid government. I finally had to hire a customs broker, who used the Canada-US connection through NATO to get release of the instruments. They were, of course, to be returned to Canada upon completion of our Arctic project.

Figure 35 - USCGC *Storis*, 1958

It was dreadfully hot on board *Storis*, so we naturally graduated to the officers' club nearby, which provided solace in air-conditioned comfort, fortified with cool drinks and the civilized company of attractive women. However, all too soon we were ready to depart Puget Sound bound for Kodiak. *Storis* was about two hundred and twenty feet in length, diesel powered, and with a small landing craft located on her forward working deck. She carried a total complement of one hundred twenty souls, which included the helicopter pilots and crew. The helicopters operated from the landing deck located immediately abaft the funnel, which abutted a small hangar.

It was a very pleasant early evening when we headed north up Puget Sound and turned westerly into Juan de Fuca Strait. The flat calm water slowly changed into a gentle pacific swell as we proceeded on our way out into the ocean. In the wardroom, a number of the ship's officers were playing poker with some concentration but, as the swell increased, the conversation slackened, cigar smoke weakened, and several players slunk off, one in particular the ship's doctor, who did not reappear in the wardroom again until we docked in Kodiak several days later.

The weather crossing the Gulf of Alaska was not too bad and was easily handled by most of the crew on board and, of course, us two Canadians. We took the opportunity to meet our American shipmates, who were the usual mix of any ship's crew—good, bad, and indifferent. The captain of *Storis* was Commander Foster, who was pleasant, but nevertheless regarded us with some suspicion. His deputy was Lt. Commander Boxroyd, who was friendlier to us and ran the deck crew with great efficiency. My cabin mate's name was Carmines, a volatile chap from Virginia, who turned out to be a good companion. The 1st Lieutenant was called West, and was the only officer on board with a claim to being of British stock, for which sin he was often teased by his fellow officers. The ship's doctor was an interesting man when he was not suffering from seasickness, and provided me with a foil when engaged in discussions on international affairs. Most other officers did not express any political opinions, even if they had them. With the exception of the doctor, they were ill informed about the world outside the USA, particularly about Canada, their next door neighbour. They were, however, a very patriotic bunch, as was illustrated at a party that took place at Boxroyd's house in Kodiak. As the party reached toward its climax, they sang the Star-Spangled Banner and God Bless America with great gusto, until I silenced them all by my rendition of Land of Hope and Glory.

Storis was secured in the naval base on Kodiak Island, a base built largely to combat the Japanese presence in the North Pacific Ocean during WW II. It was still an active base of operations in the late fifties, mainly concerned with the containment of the Soviet Union across the Bering Strait. To the northward lay the fishing port of Kodiak and the settlement containing most of the island's population. The island was fairly large and mostly mountainous in the interior, where the Kodiak bear, largest in the world, could still be found. I was afforded a close-up view of these magnificent animals when taken up into the mountains on a helicopter training trip. Our buzzing around infuriated one huge beast in particular, who rose up on its hind legs and roared at us with all the anger and disdain that it could muster.

The head helicopter pilot was a chap called Moser. He was extremely efficient and dedicated to his job and, like many people I met in the US Forces, a believing, practicing Christian. The patriot/believer combination held by many on board was a surprise to me, a product of European cynicism. Their belief shone bright and clear, a tonic for an outsider to consider and enjoy.

Oro and I were kitted out in survival suits—poopie suits to the initiated—in preparation for arctic operations, and we received aerial cameras to record ice conditions as spotted from the air. My poopie suit was so tight that I resolved to go on a diet forthwith! We prepared for our hydrographic activities without knowing in advance which areas would be ice free and available for the deployment of our equipment. The gangway guard structure was worse than useless for plotting and planning purposes, and was used mainly for the storage of the many documents our Ottawa headquarters had provided.

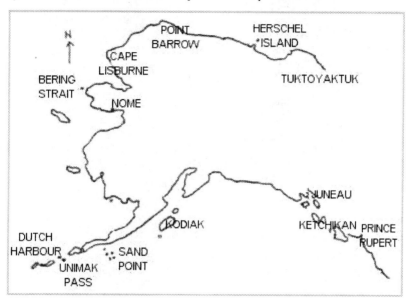

Figure 36 - Voyage to the Arctic, 1958

Our passage from Kodiak to the Bering Sea and Strait accompanied by the USS *Burton Island*, a large icebreaker, was accomplished without incident. As we moved north up the Chukchi Sea toward Point Barrow in mid-July, we began to encounter broken pack ice with the occasional large floe populated by walrus, with a huge male in the centre of the floe, monarch of all he surveyed. Sometimes, our passage rocked the floe and caused a bellowing bail-out by the disgruntled occupants. We rounded Point Barrow fairly easily but soon began to run into heavier concentrations of ice. By evening, we were making very slow progress, and a thick fog was severely cutting down our span of visibility. It then took us the best part of three days to break out of the pack beyond Barter Island and into the warmer waters off the Mackenzie River delta. At this point *Burton Island* departed to engage in other activities. Our only breaks with routine were helicopter ice reconnaissance when the weather was clear and the exciting spotting of polar bears out on the pack ice.

We had been conscious of an increasing stream of radio communication traffic as we approached Canadian territory. The Americans found it difficult at first to understand these foreign voices, but soon got used to the colourful language and phrases of the Newfoundlanders and Nova Scotians manning the DEW Line resupply vessels. This was the first year that the fleet would operate from a Tuktoyaktuk base under the Canadian flag. Previously, the DEW Line resupply had been conducted by US vessels penetrating the Canadian Western Arctic from bases in the continental USA. The cargoes were now shipped to ports on Great Slave Lake and transported by tug and barge down the Mackenzie River to the port of Tuktoyaktuk on the Arctic Ocean. *Storis'* job was to provide icebreaking support to these DEW Line resupply ships making their way through the Western Arctic to DEW Line sites stretching from the Beaufort Sea, through Dolphin & Union Strait to Coronation Gulf, and Simpson Strait to Spence Bay on the Boothia Peninsula.

Figure 37 - Seasonal route through the Western Arctic

We crossed the Beaufort Sea in light ice and made visual contact with the Canadian fleet of LSTs (Landing Ship Tanks) and small tankers, which were attempting to battle their way through the heavy ice in Dolphin & Union Strait. The ice was plugged tight in the strait, but *Storis* was able to finally break the fleet free to proceed along Coronation Gulf to their various destinations. We then made our way to Cambridge Bay, the site of a major DEW Line operation, where we took on fresh supplies and water. Using the choppers, we attempted to extend the horizontal control network out of Cambridge Bay and to run check sounding lines using the ship's landing craft. It was an abortive effort, as *Storis* had to depart for the Hat Island, Requisite Channel, Simpson Strait area, where supply ships were encountering heavy ice under pressure, being pushed south down Victoria Strait. This was a tricky area, as it had been incompletely surveyed by US naval hydrographers in previous years, due mainly to the same heavy ice conditions now prevailing. We were able to assist the supply vessels through the ice, but each of us grounded lightly while doing so. Altogether a horrible area, with numerous ice-covered hazards. We were only able to tighten up the horizontal control somewhat due to the availability of our choppers and the deployment of our tellurometers.

It was now mid-August and there was a lot of time to get back around Point Barrow before the pack moved in tightly onto the north coast of Alaska. We could have spent time usefully carrying out hydrographic projects, while providing the supply vessels with a sense of security in this hazardous area. Instead, our captain let his fears dictate his actions and withdrew his support as he hastened *Storis* westward to clear Point Barrow. On the way, we did manage to conduct sounding operations for one day off the harbour of a DEW Line site in Amundsen Gulf while the crew were given a break ashore, but then it became all speed ahead to vacate the Arctic. It was too bad—we could easily have spent several more weeks carrying out useful work. On August 28, in 5/10ths ice, *Storis* rendezvoused with *Burton Island* again off Point Barrow before departing the Arctic.

En route to Kodiak, we were deployed to examine a fleet of Japanese fishing vessels and factory ships operating in the Bristol Bay area of the Bering Sea. The fleet were catching Alaskan salmon and may well have been illegally fishing. I thought we would board one of the craft for inspection, but we did not do so because weather conditions were rough. I was surprised when we did not take boarding action, as I knew that the weather was not prohibitively bad. When I thought back to my days in cable ships and, more recently, in hydrographic launches, I was puzzled at the inaction.

We carried on to Kodiak without further ado and, much to my surprise, we were greeted by a welcoming party and a full brass band. After six or seven weeks away in the Arctic it all seemed excessive, but compared to our own underplayed and under-appreciated return to Canada a few days later, it was a well-deserved gesture. We were supposed to take passage in a week or so in a US transport ship to Seattle, but thanks to the incredible generosity often displayed by Americans to their friends and allies, we got home very much sooner. Naturally, we were spending some time in the officers' club, leering at unavailable women and drinking copious amounts of alcohol. At one of these sessions, our helicopter pilots introduced us to a P2V bomber pilot who was leaving shortly for the naval air station on Whidbey Island, north of Seattle and very close to Victoria. In casual conversation, the P2V pilot offered to transport us and our gear and equipment the following day south to Whidbey Island, if we appeared at the airstrip the following morning. We accepted his kind offer with considerable joy, countermingled with a faint sense of disbelief, and carried on with our party long after he had left.

The next morning friends from *Storis*, who had acquired a small truck especially for the occasion, helped us transport our equipment to the airstrip. There, before our still slightly unbelieving eyes, was our transport, a medium bomber converted to patrol duties, engines fired up and ready to go, with a proud pilot and crew ready to greet us. We took off swiftly and soon headed southward toward the northern tip of Vancouver Island. I had pride of place in the bomb aimers' slot and lay on my belly scanning all that could be seen below. Mostly it was a wind-freshened sea with, from time to time, fishing vessels engaged in long-line operations. Then, once we reached Vancouver Island, it became a marvelous panoramic view of the mountains and the lakes. After our slow progress on *Storis* it was wonderful to cover the miles so rapidly. Soon, we were making our approach to the Whidbey Island Air Station while viewing glimpses of the Saanich Peninsula in the distance.

Much to our delight, our pilot requested permission to transport us directly to the Patricia Bay Airport, which was granted. So, into Pat Bay roared our bomber, much to the consternation of the airport authorities and Canada Customs and Immigration. We said a hasty farewell to our aviator friends and found ourselves and our equipment surrounded by a perplexed bunch of Customs agents as they contemplated us and our South African–manufactured tellurometers. In the meantime, I was able to phone my wife, who arrived shortly afterward in our old Austin A40 together with two tiny children. She was very pregnant with number-three child and obviously labouring under the unusual heat wave still affecting our area, but still managed to ensure that Marvin and I, and our essential personal pieces, fitted into the car together with the children, who were delighted by our arrival but fast in need of cool air and cold liquids. We were almost home.

Bob Young, my boss, was most surprised to learn of our airborne landing. After a few days' leave, he arranged for us to complete the processing of our Arctic data so that we could send it on to Ottawa for cartographic corrections to the Western Arctic Chart Folio where considered necessary. He was then anxious for us to return to Ganges on Saltspring Island to complete the survey we had started in the spring.

By the time we returned to Ganges, the weather had cooled, and it had become a wonderful fall with all its glorious colours and profusion of wildlife. The harbour and approaches were thoroughly checked for discrepancies in data, and re-examined thoroughly where considered necessary. Finally, in mid-October, with the baseball World Series coming to an end, we withdrew to our Victoria base. It had been quite a survey season—from bucolic Saltspring to the High Arctic and return—with international interaction and I think mutual respect. A year to remember!

On November 1st, my wife gave birth to our second son, Duncan, who almost arrived in the Hudson Bay Company's parking lot, but managed to break an early arrival record at St. Joseph's Hospital. We were now living in Cadboro Bay, having bought a building lot and arranged for a builder to construct our new home. We were mortgaged to the hilt, but with three little children and probably more to come, we needed the extra space. Additionally, the area was grand, with large fir and arbutus trees all around, and the school a short tumble down the hill when the children would be ready.

Chapter Fifteen

The Ottawa Scene – Fulford Harbour – *Storis* Again – A Sad Ending to the Year

In March 1959, I was summoned to CHS headquarters in Ottawa to review and comment upon my 1958 data gathering voyage to the Western Arctic on board the US Coastguard Cutter *Storis*, and to plan for a similar expedition in 1959. I was booked into the elegant hotel Château Laurier where, in those days, the bed was nicely prepared each night by a supervisor and one could relax in perfect comfort and quiet while contemplating the wintry vista outdoors. It was a far cry from some of the places where I had laid down my head in recent years.

The Canadian Hydrographic Service headquarters in Ottawa were not quite up to the standard of the Chateau Laurier; indeed they were quite dowdy and rundown, being located in Number 8 Temporary Building on Carling Avenue. There was a lot of modern cartographic equipment on display and many keen young surveyors and cartographers available to explain things to a slightly bewildered person from that far-away Lotus Land on the Pacific Coast. I met some of the hydrographers-in-charge of Arctic operations, but spent most of my time with Colin Martin, the Superintendent of Charts and Acting Assistant Dominion Hydrographer, and his assistant Cyril Champ. Martin gave the broader picture of the Arctic surveying and cartographic thrust, while Champ provided me with the details and the data that would enable me to plan properly for the 1959 survey season. I also met many members of the cartographic staff, who were most helpful in advising me of their particular expertise and how it fitted into my plans.

I met Norman Gray, the Dominion Hydrographer, who seemed a very pleasant man. He did not impart any vision of where the CHS was headed in the future. Perhaps this was because, like others in senior positions, he had endured the Depression years. My final orders were drafted and I was ready to fly home.

However, before departing, I was invited to attend the annual CHS dinner and dance at the Fairgrounds, which turned out to be a great affair. I was seated at the table of Mike Bolton, hydrographer-in-charge of Hudson Bay surveys, and a number of his staff and, of course, assorted wives and sweethearts. Mike enjoyed a party, as did the other members of his staff. I particularly remember Adam Kerr, a mariner who did a hilarious take on a German oompah-pah band he had encountered in Hamburg, and Austin Quirk, who knew of some ships I had voyaged with. It was a merry night that ended all too soon with my departure to my hotel and then, far too early, to the Ottawa Airport.

My southern survey area was Saltspring Island again, with Fulford Harbour and approaches the primary area of concern. Marvin Oro had resigned to go and work for Shell Oil in Alberta, so I had a new assistant, Bob Weinberg. Bob was a very different type of man from Marvin, being a man full of ideas about everything under the sun, in contrast to his predecessor's predominant interest in the project at hand. Nevertheless, we soon worked out an understanding that enabled and enhanced the tasks before us. His different mindset, when applied to our work, influenced me in broadening my perception of events to include matters beyond sounding mileage and coastline examined. He was well educated in English literature and loved to talk about the classics, which I found interesting but disconcerting when I was obsessed with survey results. He was a local boy whose father was a mariner but now piloted vessels along the BC coast. I was sure that he would fit in without any bother.

We departed Victoria in mid-April on board our old double-ender covered wagon. The crew were all from Saltspring Island, with varying degrees of experience. I had been assured by the depot staff that our engine had been thoroughly overhauled during the winter months, but kept my fingers crossed. Our destination was a holiday camp on Saltspring Island to the east of the Fulford Harbour entrance, known as "Solimar." The camp was owned by a retired naturopath doctor, who lived away from the camp, and managed by a formidable lady who, thank goodness, was also an excellent cook. There was a small float for securing our vessel, located in a small, partially protected cove. As the holiday season had yet to begin, we were able to make full use of the recreation hall to spread our equipment around and to convert the table tennis tables into drafting tables for our field sheets. The accommodation was not luxurious, but adequate for our purpose.

Our main task, the survey of Fulford Harbour, went forward fairly well. Two unusual incidents spring to mind. The first one was when sounding the area east of Solimar, when we spotted what turned out to be an RCN training torpedo up on the rocks of a small bay. It was difficult to see at first, as its yellow markings blended in with the surrounding rocks and, intent on my sounding tasks, I was at first inclined just to note its position and pass on. However, Weinberg's intense curiosity communicated itself to me and we took immediate steps to inform the authorities of our strange find. The following day, we were besieged by the RCN in the form of a minesweeper, a helicopter, and a truck—all competing for the honour of conveying the torpedo back to the naval base at Esquimalt. I believe that we eventually received a small reward for our diligence.

The second affair was much more serious and could have led to the sinking of our launch. In the vicinity of Jackson Rock, we hit bottom while on a sounding line, and ripped all four cooling pipes out of their normal position below the hull. With water spouting up in great plumes over us all, we raced for the closest shore line. By sheer luck, we were near a nice sandy beach, which provided us with an emergency refuge. Repair equipment was obtained from "Solimar" and by high water, close to midnight, we had effected repair by the construction of cement boxes and floated off high and dry. I have always been a bit of a gambler when navigating in hazardous waters and running sounding lines over shoal areas close to the low water period. This time, my pushing of the limits had almost brought disaster!

An additional task was to conduct low water surveys in Active Pass, as a follow-up to the major survey of Active Pass and its approaches the previous year. This was a very necessary job, but one that made little economic sense, as our 6-knot vessel struggled with the ebb and flow in the pass itself and traveled the considerable distance each day from our base at Solimar. These were long days.

I once received a message from Bob Young in Victoria, directing me to go across Satellite Channel to the vicinity of Moses Point at the entrance to Deep Cove on the Saanich Peninsula, where an angry landowner was concerned about a whitewash mark our survey crew had left on the rocky point. The message instructed me to proceed immediately to rectify matters, and was received just as we were sitting down to supper at Solimar. It was a long slow journey across the channel after a hard day's work, but eventually we found the irate owner who invited us in for a drink. He lived in a large house that was filled with oriental "objects d'art," which represented his acquisitions over the years when he was employed by the Chinese Maritime Customs. He had been told that our whitewash mark was an indication that the provincial government intended to build a ferry dock on his land, which explained his concern. I assured him that we were only interested in laying down horizontal control in Satellite Channel for hydrographic purposes and would take steps to clean off the whitewash when we had completed our survey work. He was well satisfied, and we wended our long way home to Solimar.

Our coxswain was a friendly young chap named Bill Kirk who was born in Tanganyika, the former colony of German East Africa administered on behalf of the League of Nations by Britain, and now the independent country of Tanzania. He lived with his parents on Saltspring Island, where they had retired after many years in government service in East Africa. I enjoyed meeting with them and sharing experiences over a noggin or two of home-brewed beer. Our mechanic and seaman were Natives from the island and, while good workers when on the job, had a tendency to disappear for a day or so on matters of their own concern. They eventually had to be replaced by less experienced but more reliable workers.

My old grey Austin A40 came in handy for emergency transportation of spare parts and equipment. It also served as our transport when we needed to go to Ganges for a few beers. We always used the back roads to avoid giving offence to the local gendarme, which resulted in some exciting driving on those dirt-surface rough tracks through the forested hills. More than once we had to manhandle the A40 out of the salal to put it back on the track. Luckily, the vehicle was fairly light, allowing Bob Weinberg to do most of the heavy work while I attempted to navigate a path to safety. Another hazard to be avoided while based at Solimar was the excellent garlic bread produced by the manager/cook. She was a fine cook and introduced me to garlic bread during the days that I stayed at the camp to process and ink in sounding data on the rough sheets. I loved her bread and could not understand at first why I got an upset stomach every afternoon after lunch.

Even so, I became an addict, and after all these years still recall her garlic bread with fond memories.

It was now early June, and we had to return to Victoria to prepare for our Arctic season. Seattle was once more our destination, with *Storis* our base of operations. It was still an inadequate solution to our problems, but no Canadian vessel was available, so we had to make do. We were made welcome by some of our shipmates from the previous year and met the new individuals that took their place. Of particular note were the helicopter pilots, all different from those on *Storis* in 1958. My new cabin mate was Lt. Rybacki, a fine chap whose main responsibility was keeping the engines running and in tiptop shape.

We wrestled as before with the inadequate plotting and storage space provided. *Storis* departed Seattle on the June 18, traveling this time up the Inside Passage to

Juneau, before heading across the Gulf of Alaska to Kodiak Island. While the crew took leave with their families, we explored parts of the island, took part in fishing and hunting expeditions with our new friends, and even took part in a baseball game or two. The officers' club was also well patronized, with Bob Weinberg impressing the ladies with his knowledge of Dostoyevsky and the Brothers Karamazov. I think that they were more impressed with his physical attributes, but he did not seem to notice. At the entrance to the club was a huge stuffed Kodiak bear. On our visit in 1958, the bear had been in fine shape but had suffered damage since then, rumoured to be the work of inebriated pilots from a visiting Royal Canadian Air Force squadron. Our reputation was in tatters!

In mid-July, we deployed into the Western Arctic, escorted by the USS *Staten Island*, a large fleet icebreaker. Our task was to assist the Northern Transportation Company Ltd.'s fleet in their annual DEW Line resupply mission. The ice situation was much worse than in 1958, and it took more than a week to reach Barter Island from Point Barrow. *Storis* then proceeded on her own across the Beaufort Sea and into Amundsen Gulf, where she found the resupply fleet unable to penetrate the thick ice blocking Dolphin & Union Strait. The fleet anchored in Pearce Point Harbour while *Storis* carried out aerial reconnaissance of the strait and chased polar bears swimming from floe to floe.

Time was marching on in the short Arctic summer season and, in early August, *Staten Island* returned to assist *Storis* and the fleet through the continuing heavy ice concentrations in Dolphin & Union Strait. On August 16, we reached the approaches to Cambridge Bay and were able to conduct hydrographic operations east of the Finlayson Islands. By August 26, we were operating west of Cambridge Bay around Requisite Channel and Simpson Strait in relatively clear water, but inevitably the wind direction changed and polar ice came barreling down Victoria Strait, making operations hazardous. Thick fog also did not help. The resupply having now been completed east of Simpson Strait, we were able to withdraw to Cambridge Bay and then further westward. We had nevertheless achieved a fair amount of useful hydrographic surveying, both horizontal control and sounding coverage in dangerous areas, and were well content. We departed Cambridge Bay for Amundsen Gulf on September 3.

There was then one very welcome break for all hands. *Storis* stopped off at Pierce Point just north of Dolphin & Union Strait. It had a nice little harbour leading to a gentle beach in front of a small manned DEW Line site. There, the crew were allowed ashore to frolic on the beach. Beer and pizza were made available while informal games were underway and ambitious types attempted to find fish. The bay and its approaches had been fairly thoroughly surveyed by the US Navy back in 1957, but we did a bit of checking and found everything to be as anticipated. That evening, we headed westward again in a sort of controlled panic to clear Point Barrow before the ice closed in for the winter. As it was still early September and *Storis* had the larger icebreaker USS *Staten Island* to call upon in an emergency, it seemed to me that we withdrew from the Arctic with undue haste.

On September 6, we rendezvoused with *Staten Island* off Point Barrow and the USS *North Wind* off Icy Cape, and then we were truly on our way out of the Arctic.

I had several chats with the captain on the voyage back to Kodiak and found some of his views to be rather perplexing. When discussing WW II, he was quite resentful of the British, complaining that Britain still owed the US huge amounts of money from the Depression and that, during the war, US ships were charged a landing fee when docking in Northern Ireland ports. I noted that debt repayment related to WW I had been finally cancelled by all countries before it beggared the entire global monetary system, and that I had no knowledge of the Northern Ireland affair. His ancestors came from Ireland, and that may have explained his views. Most officers and crew on board were friendly to the two Canadians on board, but I found that their understanding of world affairs was not very profound, and their knowledge of the struggles between Canada and the US for control of the Great Lakes and the St. Lawrence River valley was pathetic and in some cases quite wrong. Naturally, I did my level best to educate them.

Our stay in Kodiak was quite short, but we celebrated our return by joining our helicopter pilots in the officers' club and consuming the largest steaks I have ever seen. The pilots had been our boon companions since leaving Point Barrow, as both groups had precious little to do. One of our occupations was watching movies at any hour of the day or night. That was when I learned that a Western movie was also known as a "shit kicker"—rude but descriptive. At least one of the pilots ended up in Vietnam a few years later. I hope he survived!

We were unable to repeat our 1958 experience of flying down from Alaska in a P2V bomber, but were able to get on board a commercial aircraft that flew us into Seattle. We were home in Victoria on September 16. Our equipment followed us courtesy of *Staten Island*. I rejoined my growing family before returning to Fulford Harbour to complete our survey work. In mid-October we returned to base just before the arrival of my father and stepmother all the way from Scotland. It was grand to have them with us, but it placed quite a strain on my wife, who was already having a hard time of it looking after our three tiny children. I did not realize that she was close to a breakdown and probably did not help in any significant way. She did, however, carry on, and my parents were well looked after during their stay. Additionally, our friendly neighbours made sure that they were kept entertained. In mid-November they left, bound for Ontario and Quebec to visit other friends and relatives before returning to Scotland.

In late November, I was ordered to go to Ottawa to submit our Arctic data for inclusion on the temporary charts being prepared for the 1960 season in the Arctic. Shortly after my arrival, I was advised by my wife that my father was in hospital in Montreal with a torn esophagus and was very ill. I was able to go to Montreal several times to see him, but I could do little to help him. He was unhappy so far from home and—I believe—wanted to die. He even attempted to remove his life support system in my presence on one occasion. He died on December 19, 1959, and I was bereft. He was a fine, upstanding man, a man of integrity and fun. He would have so enjoyed telling his many friends of his adventures in Canada. Alas, it was not to be. It took me quite some time to mourn him and to accept his passing.

Chapter Sixteen

A Canadian Presence in the Western Arctic

Our reliance on the US in our own waters had to come to an end, and in the autumn of 1959 the brand new Canadian Coast Guard Ship *Camsell* was delivered to the Victoria base, preparatory to her employment in the Arctic. In the meantime, she would be employed as a buoy tender and lighthouse resupply ship on the coast of British Columbia.

Figure 38 - CCGC *Camsell*, 1960

Camsell was about two hundred and twenty-three feet in length, had a forty-eight-foot beam and a draught of sixteen feet, with raised forecastle and heavy-lift derricks on the foredeck to deploy over the working deck and storage holds. Abaft the working deck lay the main housing, surmounted by the bridge and monkeys' island. Immediately aft of the bridge deck, one thick funnel protruded skyward, with a hangar and helicopter landing deck running aft to the stern of the ship. All the ship's accommodation was contained within this after-housing. *Camsell* was powered by twin diesel electric engines and could cruise comfortably at 12 knots. She displaced 2,022 gross tons. For BC coastal operations, she was fitted with a regular workboat located on her foredeck. For hydrographic operations, she would be fitted with a twenty-six foot hydrographic launch. I was well pleased with events and left for my annual planning trip to Ottawa in good spirits.

After completing my planning work with the assistance of Cyril Champ and Colin Martin, I was free for a short period to absorb the Ottawa political scene. Back in 1957, a Conservative government had taken over from the long-time Liberal governing party. It was in a minority position in the House of Commons, but the following year held another election in which they swept the country. I was happy with this event, and decided to visit my own MP in the House to attempt to find out how well the new government was conducting the business of the nation. He was a chap called Albert McPhillips, a lawyer of course, who seemed quite pleased to see me, and gave me a tour of Parliament before treating me to lunch in the Members' dining room at the House of Commons. I enjoyed it all thoroughly, particularly the wonderful circular library, and came away wistfully wishing that I could take part in the political scene. I did not realize it at the time, but the Diefenbaker government was probably at the peak of its power and influence and was destined to eventual defeat in a few more years.

Upon returning home to Victoria, I decided to take a look at our BC government but was unable to penetrate the halls of political power. We were ruled by a peculiar group known as Social Credit, an amalgam of former Tories and Liberals united in their desire to keep the Socialists from power. Our Premier was W.A.C. Bennett, a former conservative who had won a slim majority and appeared to have his finger firmly on the political pulse of the BC voter. He had done things not expected of a conservative, by provincializing the BC ferry system and by absorbing the large BC Electric Company into a provincial power corporation. By doing so, he gave the province a proper ferry system, and he met the power demands of an expanding provincial economy by investing in huge dams in the interior. He was accused of juggling the finances of the province, and no one ever confessed to voting for him, but he remained in power. A truly remarkable man.

I had now become aware of the officers that I would have to deal with on *Camsell*. First of all there was of course, Art Davidson, the Master. He was originally from Ontario, but had grown up on the Pacific Coast. He had spent many years in command on Coast Guard vessels on the Pacific Coast and he was very knowledgeable on coastal matters and in seamanship skills. He was not a large man, but reflected an aura of decisiveness and competence. He was interested in what I hoped to accomplish in the Arctic. He was impulsive, as I tended to be at times. I felt that he and I could get along very well. John Strand was his chief officer, a quieter but very competent man, probably an excellent foil for our captain. One of the junior officers was a young Englishman, Peter Stocker, still learning his trade. The chief engineer was a thoroughly capable individual who kept the mechanical and electrical side of things humming along.

Meanwhile, Federal Electric, the company operating the DEW Line, had agreed to provide accommodation, victuals, and transport to the hydrographic advance party that would precede *Camsell* into the Western Arctic and conduct horizontal control surveys at the beaching areas of a number of key DEW Line sites. This was a most welcome development and greatly appreciated.

Weinberg and I departed Victoria in mid-June, bound for Edmonton and Cambridge Bay prior to rendezvousing with *Camsell* in the Western Arctic in July. In Edmonton, just before I boarded our flight north, I suddenly realized that I did not have a towel, so purloined one from the motel where we resided. The towel became a bone of contention over the following months, as the motel management tried to recover its loss from my poor wife at home, who refused to accept that her husband could be guilty of such a dastardly deed. The matter was finally settled to everyone's satisfaction. Our flight was filled with wild construction workers on their way north to make their fortunes, and fresh groceries of every description. Unfortunately, our plane developed engine trouble after about one hour's flight out of Edmonton, and we had to jettison fuel before returning. After several hours' delay we were on our way again. At Cambridge Bay, we stayed in a tent close to the large DEW Line site, where we were able to partake of meals and showers, while we carried out some basic horizontal control work with theodolites and tellurometers. Upon completion of that task, we flew eastward to other DEW Line sites at Shepherd Bay, McLintock Bay, and Jenny Lind Island, where we conducted similar exercises. We then moved westward, touching down at Lady Franklin Point, Tysoe Point, and Pearce Point harbour before completing our advance task at Cape Parry, where we rendezvoused with *Camsell* on August 9. We were then transported on board by the ship's helicopter.

Camsell proceeded directly toward Queen Maud Gulf in mostly light ice conditions and commenced laying navigational buoys around Hat Island and in the western approaches to Simpson Strait. Additionally, very useful track sounding was carried out in previously uncharted waters. Unfortunately, our hydrographic launch *Quail* was not available for sounding operations due to engine problems, but we were able to make very good use of the *Camsell* helicopter, which allowed us to complete a horizontal control network tying in Jenny Lind Island through Simpson Strait to Gjoa Haven and Shepherd Bay. These theodolites and tellurometers were worth their weight in gold. Our pilot was Bob Masters, a former RAF bomber pilot, who had been shot down over Germany on his third sortie. He had fascinating tales to tell of life in a POW camp in northern Germany.

We had ample time to use the helicopter for ice spotting and general reconnaissance work. Bob cheerfully flew us wherever we wanted to go, keeping an eye out for roaming wildlife and signs of Inuit presence in the past, such as rock cairns, and generally acting as an unpaid tour guide.

Camsell was not required for direct icebreaking duties in support of the DEW Line resupply due to her late arrival in the Beaufort Sea. However, as previously noted, she did sterling work in Queen Maud Gulf eastward in the laying of navigational buoys and the construction of radar reflector towers. She had badly damaged one of her propellers in breaking her way along the Alaskan shore from Point Barrow to Barter Island, and so sought temporary refuge in the harbour at Gjoa Haven on King William Island to attempt to change propellers. Her first attempt at beaching and canting the vessel to effect the change was a failure, so a call went out to the RCN, which quickly provided a diving team who successfully completed the project. Gjoa Haven was of course the harbour where Amundsen wintered with his wooden hulled *Gjoa*, when he made the very first circumnavigation of the Arctic in 1905–1906.

We then departed for Cambridge Bay to take on fresh water and food supplies. The main DEW Line site was visited—a remarkable technical achievement situated above the Arctic Circle. The site stood above the permafrost on gravel embankments and consisted of many prefabricated units joined together to provide heated living quarters and operational areas for the radar operators and communication experts who manned the site. The flags of Canada and the United States flew over the site, symbolizing the joint commitment to our defence of North America. Everything was spotless and efficient and even comfortable, and maintained in that fashion. The food was of high quality, with a rotating three-week menu. It was possible to phone directly to anywhere in North America from the site, and one could also get an on-the-spot weather forecast from other sites along the line. An impressive achievement, and a prod to our own Department of Northern Affairs to modernize other government structures and organizations in Cambridge Bay and other settlements in the north.

By now, it was early September, and as all seemed well with the resupply fleet, we headed westward in moderate seas before encountering 5/10 ice along the shoreline from Barter Island to Point Barrow. South of Point Barrow, we provided assistance to *Mohawk*, an NTCL tug with crippled propellers. Finally, we towed her all the way through the Bering Strait and the Bering Sea to Dutch Harbour, where proper repair facilities were available.

It was rather stormy in the Bering Sea, but the tow was secure, and we made the haven of Dutch Harbour in good time. Dutch Harbour was an inactive former US naval base of great importance in the Pacific Ocean struggles of WW II. The base was situated on Amakanak Island, which is today connected to the main island of Unalaska Island and the port of Unalaska by a bridge. Unalaska Island is part of the Fox Islands group of the Aleutian Islands, lying immediately west of Unimak Pass, the normal entry point into the Bering Sea. Although Dutch Harbour itself was largely deserted in 1960, the port of Unalaska was busy with fishing activities.

The wharves at Dutch Harbour were still in good shape, and we tied up close to the US Coast & Geodetic Survey ship *Rainier* out of Seattle. We had a useful exchange of hydrographic information with officers of *Rainier*, and entertained one another on board our vessels. Fresh water was available off the wharf, and fish and other fresh food was transported to us from Unalaska. Meanwhile, we explored the ruins of the base, finding odd items such as terrazzo flooring in fairly good shape after fifteen years of peace, and an underground passage that led to the base command room located below the signal tower. On the walls of the command room were the fraying charts and maps depicting the status of the war in the Pacific on the day of the Japanese surrender in 1945.

We returned to Victoria on September 13 in good order and were greeted by several dignitaries, and our wives and families. No brass band, but some attention from the press and lots of love and kisses from the ladies and our offspring. It was nice to be home, although we were soon packed off to assist other hydrographic operations prior to the winter break. On November 23, our third son Malcolm arrived on the scene. We were fast becoming a clan of our own. A number of people made life easier for my wife and me during this period of enlargement of our family. My wife's parents and good friends like Gordon and Betty Schenk and Ralph and Sue Wills particularly spring to mind.

I then headed back to Ottawa for the review of our 1960 work and preliminary planning for the coming 1961 season. I made the case as strongly as I could that, while *Camsell* was a big improvement over *Storis*, her primary mission was not hydrography. We needed our own dedicated vessel as soon as possible. Various approaches were discussed, but all suggestions were of an ad hoc nature and lacked a real commitment to the support required for a modern hydrographic survey in the far north. Without adequate technical backup, we were whistling in the wind. It was an illustration to me of how much CHS management was still in thrall to the Depression era. I returned to the West Coast a little perturbed about what the 1961 season might bring.

Chapter Seventeen

A Dubious Venture – A Lesson Learned

In January of 1961, I travelled to Edmonton to take part in a planning workshop sponsored by the Northern Transportation Co. Ltd. to discuss and decide on an operational plan for the 1961 season. Invitations had also been sent to representatives of other agencies with an interest in Western Arctic affairs. It was very cold, even the hotel seemed unable to combat the sub-zero temperatures, and walking up Jasper Avenue with a paralyzing northerly wind was enough to put anyone off further ventures in the Arctic.

We did, however, survive the elements and had a very useful meeting, where I received a promise of NTCL assistance when the CHS would base a hydrographic vessel at Tuktoyaktuk, their supply port at the mouth of the Mackenzie River. Additionally, contacts were made with several influential individuals, which proved of great use in the years ahead. I also committed CHS to a larger role in the charting of Arctic waters, to meet the demands of not only the DEW Line resupply but also the anticipated general increase in trade as the Western Arctic opened up to further exploration and exploitation of its natural resources.

I returned to Victoria to unfreeze and to prepare for a final planning session at CHS headquarters in Ottawa. Upon arrival there, I found that management was keen to expand our efforts in the Western Arctic, suggesting that, in addition to manning *Camsell* again on an opportunity basis, I consider transporting the CSL *Rae* from Great Slave Lake up the river into the Arctic Ocean and along the Arctic channels to Cambridge Bay. I was taken aback. *Rae* was an elderly, large, wooden-hulled craft with erratic power plant and no shore-based maintenance backup. I could visualize getting her to Tuktoyaktuk without too much trouble, but taking her through ice-infested waters along the Arctic coast was another matter. There were DEW Line sites along the route, but most had little shoreline protection and none had permanent dock facilities. Finally, none provided the assurance of technical repairs in emergency situations. I shuddered as I thought of all the things that could go wrong on such a voyage, and said no!

An alterative was then suggested, after a luncheon meeting between the Dominion Hydrographer and the Commissioner of the RCMP, where perhaps we might have the use of the RCMP launch *Spalding* based in Cambridge Bay. It sounded more promising, so arrangements were made for me to meet with the Superintendent of G division (Arctic) as soon as possible. On my visit, I was accompanied by Mike Bolton, who had great experience as hydrographer-in-charge of field parties in the Eastern Arctic, and was becoming a closer acquaintance as my Ottawa visits multiplied.

We reported to RCMP headquarters, a rather intimidating building and, after passing through various check points, were ushered into the august presence of the Superintendent of G division by a staff sergeant who looked half terrified while he heel-clicked his ramrod-erect self in a perfect imitation of the sort you see in old movies about the German SS. His superior was also ramrod straight, even sitting behind his desk, and greeted us very formally and without a sign of friendliness. He informed us that we could have the use of *Spalding* on an opportunity basis but that, even then, the launch would remain under the control of an RCMP officer at all times. I tried to advise him of my concern that such an arrangement was unworkable, particularly as I was a Master Mariner myself, but he emphatically stated that we either accepted the arrangement he had indicated or there was no arrangement. I was unhappy, but thanked him and advised him that I would report back to CHS headquarters my views on the operating conditions proposed. We went through another heel-clicking demonstration while being ushered from his presence. I was not unhappy to flee the building— the excess of militarism had given me a bad taste, and the terms of agreement, as proposed, were already causing me some concern.

Having turned down the *Rae* proposal, I was in a difficult position regarding the suggested RCMP arrangement. If I followed my instincts and said no, there was no other offer on hand. After some discussion with Martin and Bolton, I decided to take the RCMP offer. Plans were drawn up with the assistance of Champ for possible operations out of Cambridge Bay utilizing *Spalding*, and another set of instructions was prepared for the party operating off *Camsell*. In theory, both parties would report to me, but distance and circumstances would allow the *Camsell* hydrographers a fairly free hand.

I returned to Victoria still wondering if I had made the right decision, but determined to make the best of it. Bob Weinberg would be joined by Graham

Richardson on *Camsell*, and I would be joined on the *Spalding* party by Jack Chivas, another mariner. We departed Victoria in late June, bound for Cambridge Bay. We had no crew of our own and were advised that we would need to fend for ourselves. A short stopover in Edmonton allowed us to make arrangements for food to be flown north with us to Cambridge Bay, and to ensure that a walled tent would be made available to us for accommodation and working area. It was a dismal start to our project.

The Anglican minister kindly offered to have his wife cook one hot meal a day for us, providing we supplied the essential food. So we heathens ate one meal a day in the rectory, duly blessed by the priest. Another piece of luck was the presence, on the southern shore of Cambridge Bay, of a Surveys & Mapping geodetic party, who gave us the use of one of their helicopters from time to time to conduct reconnaissance surveys and establish horizontal control. One of the pilots was Peter Corley-Smith, a former RAF pilot with the Special Operations Executive in WW II. We became good friends, enhanced by finding that he and I lived fairly close to one another in Victoria. It was a sad day when the geodetic party moved away to another site.

The only problem with eating at the rectory was overcooked meat that arrived on the table. The meat purchased in Edmonton was prime, both steaks and roasts, so to see it always well done was a travesty of what we had been expecting. That, and a vision of Yorkshire puddings served before the meat and covered in gravy, was a real downer on a Sunday when we awaited our meal with some anticipation. However, there were compensations. The reverend, his wife, and two young daughters had fine voices and entertained us with renditions from Gilbert and Sullivan operettas. It was better than skulking in our leaky tent all day.

Spalding was still up on stocks, as the harbour was still packed with ice when we first arrived. The RCMP post was manned by a corporal, two constables, and several native auxiliary constables out on patrol. We made ourselves known to the corporal, but little enthusiasm was shown. It did not appear that any repair or refurbishing work had been carried out on *Spalding* for some time, her seams had cracked open, and she did not look entirely seaworthy. My expressions of concern brought little response other than a shrug indicating that the problem would be taken care of eventually. The two constables, however, were pleasant young fellows and quite keen to be involved in our hydrographic project.

Eventually, work commenced on repairing and sprucing up *Spalding*, and soon she was launched into open water to await sea going trials. She was a Cape Breton hulled craft, built in the Maritimes a number of years previously, and shipped down the Mackenzie River by barge, then by Hudson Bay Company freighter from Tuktoyaktuk to Cambridge Bay. She had not been deployed on any major voyage prior to our arrival, as far as I could ascertain. She was wooden hulled and about thirty-two feet in length, with a small pilot house located on the main deck in front of a cargo hold. Cabin space in the bow was accessed by a ladder from the pilot house. A gasoline engine was the means of propulsion and was situated immediately below the pilot house.

Figure 39 - RCMP launch *Spalding*, Cambridge Bay, 1961

In late July the HBC freighter *Banksland* arrived in Cambridge Bay, signalling the start of the short navigation season. It was a time of great celebration among the locals, and with the extended hours of daylight no one, including even the tiny children, seemed to go to bed. As mariners and hydrographers, we were welcomed on board *Banksland* to collect our shipment of equipment, and stayed to consume a considerable quantity of the Master's rum—A great way to forget our troubles, with only the inevitable ghastly hangover to face as payment for our overindulgence.

A few days later, we set off on our first voyage, with an RCMP constable ostensibly in command. We jury-rigged our echo sounder transducer and attempted to run a number of sounding lines south from Victoria Island into Coronation Gulf, while extending the horizontal control eastward toward Victoria Strait. It was an abortive effort, as our transducer gave us a great deal of trouble and we lost a tellurometer when our skiff capsized in the surf. We returned to Cambridge Bay to lick our wounds.

Our next sojourn to sea was at the behest of the corporal, who wanted *Spalding* to journey across Coronation Gulf to an Inuit fish camp that was supposed to be located in a group of islets close to the mainland shore. Once more, with a constable in command, we set off, I must admit with some misgivings. Our magnetic compass was not of much use in these latitudes, and our radio communications were not the best. There were no nautical charts of the mainland shore to the south of us, and little indication of any prominent features that would assist us in our endeavour. My worst fears were realized when we found ourselves in poor visibility right in the middle of a mess of islets and rocks. We snagged our rudder on a rock that was barely awash and decided to anchor while we undertook repairs and tried to work out where in heavens we were.

We were able to communicate with *Camsell*, as she was located further east in Queen Maud Gulf, and advised her of our ridiculous situation. She arrived, standing well off our maze of hazards, and provided us with navigational advice utilizing her superior height above the waterline and, of course, her several radars. We returned to Cambridge Bay duly chastened, but fuming at the command structure and the problems it created.

Our next venture was westward along Coronation Gulf toward Lady Franklin Point where the higher land made navigation easier and the DEW Line sites provided technical and other assistance.

There is nothing like a shower, good grub, and an up-to-date movie to help make one feel civilized once more. The higher land was also interesting, with its ancient sand and shells indicating a different sea level in days long past. We were now ready to undertake a longer voyage and headed for the entrance to Bathurst Inlet on the southern shore of Coronation Gulf. The land was steeper, even mountainous in places, and quite spectacular. Again, signs of higher water levels were found when exploring bays and valleys, and signs of caribou were abundant. One problem was the unfortunate habit of the RCMP officer, who placed *Spalding's* bow up on the sand or mud each night rather than anchoring offshore as was good seamanlike practice. As a result of such sloppy action, we had to spend some time every morning warping ourselves off the beach. The tidal range was small—about two feet maximum—but still enough to put us aground.

The settlement at the southern end of Bathurst Inlet was not substantial. It consisted of the Hudson Bay Company post, a tiny Catholic church, and a number of shacks/tents. Everyone was glad to see us and did their best to entertain us. The Hudson Bay factor was from the Orkney Islands and the priest from Brittany; the remainder of the population were a mix of Inuit and Indian. We spent a night there, talking and sampling rum into the early hours of the following day. Our route north out of Bathurst Inlet took us westward toward Arctic Sound, where we met a small band of Inuit who still practiced the old hard way of life of their forefathers, despising the soft existence of their fellows huddled around settlements. The group that we contacted was made up of an elderly man who was partially crippled, two women who were his daughters or daughters-in-law, and several small children. The other men in the group were apparently off hunting or fishing. These people were wonderfully tough and self-reliant. They did not really want contact with us, but finally agreed to exchange fish for a small tin of tobacco and a package of butter. The tobacco was for the old man and the butter for the children; the women wanted nothing. I would not have dared to mess with them—they were powerful hard specimens of womanhood—and looked capable of overcoming any assault upon them or their kin. We departed their company knowing that we had probably witnessed the last free independent Inuit.

We had conducted track sounding throughout our Bathurst Inlet exploration, wrestling all the while with our inadequate transducer arm. A more suitable arrangement had to be constructed prior to our next voyage—if there was to be one. We returned to Cambridge Bay in a slightly better frame of mind, feeling that our latest voyage had been a success.

A better transducer housing was built and we were allowed to deploy *Spalding* for several days in the western approaches to the bay, increasing the overall sounding coverage and delineating several shoal areas. It was our most productive time on *Spalding*, and we took full advantage of the opportunity. It was now early September, and the first snow flurries were heralding the approach of winter. So, just as success was within our grasp, we were diverted to a region where the RCMP had a camp, to collect a native constable and his small family, and to load dried fish onto *Spalding* for winter feed for the dogs.

The camp was on a small creek about one hundred yards inland from the shore, so we had to laboriously kedge our way up the creek to load our cargo and then even more laboriously kedge our way back to sea, while loaded down with the constable and his family, six huskies, a dozen dead seals, and roughly two tons of dried fish. How we made it out of the creek was a miracle, but little did I know that there was more to come.

The weather was freshening as we departed for Cambridge Bay, and increased to storm levels as it was getting dark. Accordingly, we had to find shelter immediately, and worked our way into a sheltered bay with the ominous name of

Starvation Cove. By that time I had taken command of *Spalding*, saw to the anchoring of the vessel, and belayed the RCMP's silly instruction to put the native constable, his family, and the huskies on the beach for the night. Everyone huddled in shelter, except the poor dogs that remained on deck. In the early hours of the morning, something awoke me and I discovered that the lower cabin space was fast filling with water. The extra cargo on board had set *Spalding* lower in the water, where improperly caulked seams allowed the elements to pour into the hull. It became a frantic scene of all hands to the pumps and any other means we could use to save the vessel. The engine was also flooded, so the problem was compounded. Eventually, after hours of hard pumping and bailing, we were able to start the engine. The weather outside the cove was still not good, and the RCMP constable wanted to remain at anchor, but I had lost my patience with the idiotic command situation and decided to up anchor and run for the shelter of Cambridge Bay while keeping her afloat by pumping madly with all hands taking turns. We departed Starvation Cove (christened by Collinson's sailors while on their search for the Franklin Expedition in the 1850s), bounced about a bit, but made it into the settlement without further incident.

It was now time to leave for the south and, a few days later, we managed to get on a flight to Winnipeg after saying farewell to our friends in Cambridge Bay. We were glad to be in Winnipeg, but even gladder to return to family in Victoria. It had been a memorable season in the Arctic. I was anxious to convey my thoughts to my superiors.

Fate then lent a helping hand. The director general of the Surveys & Mapping Branch, Sam Gamble, was visiting British Columbia and my director, Bob Young, arranged a meeting with him, where I was given an opportunity to expound on my concerns and to sketch a scenario that might in part meet our immediate charting requirements in the Western Arctic. He was a blunt man, with an abrupt manner of speaking, but I felt an immediate rapport—his word could be trusted. He questioned me closely about *Spalding* and the *Rae* proposal, and then asked me what would be my minimum needs if a vessel were to be made available. I hastily and verbally described a smaller version of a halibut fishing vessel and had an immediate indication that my suggestion stood a good chance of fulfillment. Upon reflection, I really should have held out for a slightly larger vessel that would perform well in extreme ice conditions and contain adequate electronic equipment and space for crew. I think I must have been seduced in part by the penny-pinching ways of management, and a tremendous desire on my part to acquire a ship of my own. The die was cast, and I would live with the outcome.

Plans were drawn up in short order for a steel-hulled vessel of sixty-six feet in length, to be built on the west coast of Canada. The contract to build the vessel went to the Star Shipyard of New Westminster in late 1961, and Bill Reid, a well-known naval architect with a lot of experience in building halibut fishing vessels, was hired to supervise construction. When I flew to Ottawa for my annual Arctic project debriefing, I was a fairly happy man.

In addition to the debriefing, I renewed friendships with Dusty DeGrasse, who had been on the *Willie J.* with me back in 1953, and with Mike Bolton, who was to become much better known to me as the years went by. I returned home and celebrated a merry Christmas with my family.

Chapter Eighteen

CSS *Richardson's* Voyage into the Western Arctic—1962

It was a very busy winter, with my hydrographic responsibilities for the Western Arctic taking second place to my concern and care about the construction of CSS *Richardson* that was taking place across the Strait of Georgia on Annacis Island in the port of New Westminster. I became a regular traveller on the BC ferry system, meeting with the shipyard officials and the naval architect fairly frequently. *Richardson* was taking shape quickly, her tight-frame spacing and after-cabin fittings the main changes to a regular small halibut-fishing vessel hull.

Figure 40 - CSS *Richardson* awaiting launch, Star Shipyard, Annacis Island, March 1962

Additionally, I wanted to ensure that she was fitted out with good radio communication equipment, a gyrocompass system, and the latest echo sounders. In these matters, I was assisted by Roy Ettershank, an ex-hydrographic surveyor. These matters I will report upon later. *Richardson* was designed to be engined with Stork diesel engines but, facing a possible shortage of Stork engine parts in the Arctic region, it was decided to go with tried and true, if noisy, General Motors engines. Otherwise, everything seemed to be straightforward, and in late March we assembled to properly launch our little ship.

Figure 41 - Dignitaries present at *Richardson* launch, March 1962

It was quite an occasion, on a bright but breezy day, with a moderate number of people in attendance at the Star Shipyard. There were, of course, the two Mercer brothers who owned the shipyard, one the yard boss and the other the financial manager, and all the shipyard workers who had helped put the hull together. The official representative of the Government of Canada, Bob Young, the director of the CHS Pacific Coast Office, and Gerry Andrews, Surveyor General of BC, representing the provincial government, were on the platform, together with Bill Reid, the naval architect, the financial officer for the Government of Canada, and other odds and sods, including me, and a varied assortment of spouses of those in attendance. The actual launching ceremony was carried out by Mavis Young, my boss's wife, who did a thorough job of smashing the champagne bottle across *Richardson*'s bow. She was afloat, and we were closer to being in business.

Figure 42 - Mavis Young successfully christens *Richardson*, 1962

The name *Richardson* was chosen to honour Sir John Richardson, a prominent explorer of the Western Arctic region in the 1820s, and the strong deputy who helped make the first Franklin expedition along the Arctic Ocean shoreline a resounding success. I hoped that the name would bestow that type of success on our endeavours.

Figure 43 - Sir John Richardson

I then went off to Ottawa for final briefing prior to our first voyage into the Arctic. In addition to *Richardson*'s plan of action on an opportunity basis, plans had to be drawn up for the *Camsell* party to work on joint projects wherever possible in the variable ice and weather conditions that would inevitably affect our program.

My visits to *Richardson* became more frequent as equipment was fitted on board and she gradually changed from a hull to a living, breathing vessel. I was satisfied with most of what went on, but was dubious about the effectiveness of our radio communication equipment, which I knew was essential to our success in journeying into the treacherous waters of the far north. The experts that I consulted seemed unable to understand how difficult it was to receive up-to-date ice and weather information while at a distance from stations such as Point

Barrow. I hoped I was wrong, but had an uneasy feeling that our communication system left something to be desired.

In late May, amid jubilation, *Richardson* was delivered for local trials, and our first crew members were hired. They were: one experienced hand, Jack Cunningham, from the CHS depot in Victoria, a keen, young, inexperienced hand called Ron Longbottom, also from Victoria, an experienced older engineer, Percy Napier, also from Victoria, and a cook named Coyle who fed us well but had a drinking problem. Other crew members would be picked up later. A senior member of the CHS ship branch arrived from Ottawa and helped smooth the way for me with the shipyard, by taking care of deficiencies in meeting in full the terms of the contract and in fulfilling the requirements of the Canada Shipping Act. I was now ready to assume my dual responsibilities as Master and hydrographer-in-charge of CSS *Richardson*.

In early June, we undertook the circumnavigation of Vancouver Island as a thorough test of the vessel's viability. Cap Brown, now in charge of sailing direction, joined me onboard as mate and to impart some of his knowledge and experience of the treacherous waters of the west coast of Vancouver Island. It was a grand voyage in almost perfect weather, and we were able to put *Richardson* through her paces. Stops were made at Campbell River and Alert Bay before rounding Cape Scott to commence the southern leg of our journey. Our first port of call on the west coast was Winter Harbour off the entrance to Quatsino Sound. By now the clear weather had turned to thick fog as we made our way southward toward the Brooks Peninsula. Our radar was working perfectly and we were able to anchor off Kyuquot in Kyuquot Sound with little difficulty.

With the fog partially clearing, our next destination was Esperanza Inlet, which we approached as the mountains inland were just clearing of cloud. Useful sailing directions information was obtained at Tahsis and Zeballos before going on to Nootka Sound and Gold River up the Muchalat Inlet. Our final visit was to Clayoquot Sound and the village of Ahousat, where we viewed the famous hot springs. Everything on board appeared to be operating efficiently, with the exception of the engines and the hydraulic steering gear, which would require adjustment back at the shipyard. We headed southeastward past Barkley Sound into Juan de Fuca Strait and a welcome return to our home port of Victoria.

We then crossed the Strait of Georgia, entering the Fraser River en route to Annacis Island and the Star Shipyard. The necessary repairs and adjustments were carried out in the yard while I welcomed Al Ages, who had been appointed my mate for the Arctic season. He and I had been on the *Willie J.* together, so that I knew what to expect of him. He was at the University of British Columbia, working on an engineering degree, but more important to me was the fact that he held a Dutch Master Mariner's Certificate. He had spent a number of years with the Dutch Merchant Marine before becoming a hydrographic surveyor, he was keen, reliable, and looking forward to our arctic adventure. What more could I ask?

It was early July when we departed Victoria with a few of our children on board for the short voyage to Clover Point. We then headed northward, but shortly thereafter our cook, who had been exhibiting signs of a gigantic hangover, announced that he was quitting and demanded to be set ashore. He was terrified when contemplating our proposed voyage into the Arctic. At this time, the government had put a temporary freeze on hiring, so we left our cook on the wharf at Nanaimo and hoped that we would get authorization to hire a new cook when we reached Prince Rupert, our last Canadian port of call before reaching the Yukon and Northwest Territories.

Jack Cunningham filled in as temporary cook and we survived until reaching Prince Rupert. The weather and passage were favourable, so that we made good time up the Inside Passage to our first destination. It was easy to obtain fuel, food, and fresh water, but very difficult to find a cook. We finally had to settle for a young lad supplied through the United Fishermen and Allied Workers' Union, who honestly professed that he was not a cook but that he would try his best to meet our needs. At least Vic looked strong enough to act as an extra hand on deck when required.

Unfortunately, his professions about cooking were all too true, and we suffered accordingly. Everyone pitched in and showed unexpected talents, with Ages turning out quite passable freshly baked bread. The journey up into Alaskan waters was not solely concerned with gastronomy; it revealed many interesting features from Ketchikan and Petersburg, past Icy Strait and the glaciers to the coastal hub of Juneau. We briefly sampled the honky-tonk pleasures of Juneau's Main Street establishments before heading seaward toward the Gulf of Alaska.

After the smooth waters of the Inside Passage in British Columbia and in Alaska, the gulf swell soon woke us up to the possible perils that we might encounter. A gigantic whale, the same length as *Richardson*, surfaced noisily just in front of us, leaving us in awe while wondering what other creature from the deep might appear before our eyes. However, the only other sights worth mentioning were the many fishing vessels south of Kodiak Island and their long lines of nets stretching for many miles on the surface of the Ocean.

Arrangements had been made with the USN Base at Kodiak for fuelling, fresh supplies, and fresh water, and they made us welcome while seeming quite amused by our size and pretensions. Our invoices for fuel alone required sixteen copies to be signed, a little bit much. I resolved to go to the commercial harbour if we ever returned to Kodiak Island. Our route from Kodiak took us west-southwestward to the entrance to Unimak Pass, the gateway into the Bering Sea. There we encountered a very confused and turbulent sea that drenched the engine room skylights and knocked out our battery power. We soon recovered and continued on our way, chastened but unbowed.

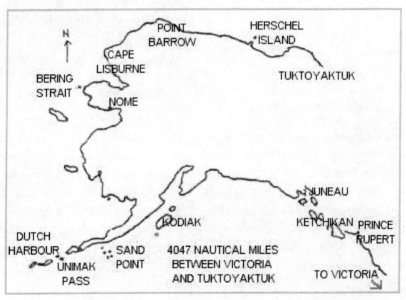

Figure 44 - The long voyage in 1962

Our next stop was the famous old gold mining town of Nome, located east of the Bering Strait and tantalizingly close to Siberia. We tried to no avail to cross the sandbar off the entrance to Nome harbour, finally giving up the attempt while wallowing at anchor close to the bar. Our workboat was launched, and a short expedition was undertaken to explore the sights and delights of the city of Nome. It seemed to consist mainly of a long, straggling street occupied by shacks of various age and description, most of which were saloons, bars, and houses of ill repute. All, however, appeared to be doing a roaring business. We chatted with the locals, who were interested in our vessel, even managing to encourage the local newspaper editor to write us up in the local newspaper. Another character wanted us to tow him across the Bering Strait to Siberia. I was interested, but decided that the political situation was too dangerous. I could see headlines in the

Canadian Press about our arrest by the Russians—no thank you! We eventually moved eastward to a sheltered bay, where we were able to take on fresh water.

We passed through the Bering Strait gazing with interest at Big Diomede and Little Diomede Islands, the symbols of Soviet and US power lying in the middle of the strait. Save for radar antennae on both islands, little else could be seen of that power. We crossed into the Chukchi Sea and started to see signs of ice about a hundred miles south of Point Barrow. There were also a number of walrus carcasses minus their tusks floating in the water, and then we found the main walrus herd floating on several large ice floes, making a great deal of noise when they saw us approaching. There were perhaps fifty or so walrus on each floe, clustered around a gigantic bull walrus that controlled his harem. At our approach, the harem scattered, plunging into the water with roars of displeasure, leaving the master of the harem to struggle out of the hole he occupied in the center of the floe while bellowing his defiance at us, intruders into his domain. It was a spectacular sight, leaving us feeling good about seeing this display of wild walruses, while saddened to see the carnage caused by indiscriminate killing of the beasts for their magnificent tusks.

In a short time, we had anchored off Point Barrow in fairly light drifting ice, and went ashore to get the latest ice and weather forecast for the north shore of Alaska between Point Barrow and Barter Island. Barrow itself was unspectacular, consisting of shack-like buildings close to the shore, with muddy, wet, and sandy trails fanning out in every direction. There were a number of tourists who had just flown in for a short visit, and a lot of local Inuit. To our vast amusement, the tourists, who seemed to be mainly from Texas, were bundled up in parkas and mukluks, while the locals sported blue jeans and baseball hats.

The ice information received being favourable, we hoisted the anchor and traversed around Point Barrow into the Arctic Ocean. All along the shoreline, the heavy ice had grounded in ten to twelve feet of water. *Richardson* drew seven feet of water so, by moving with extreme caution, we were able to make our way eastward while in constant sight of the Alaskan shoreline. Fog inhibited our progress from time to time, but our radar enabled us to make good mileage and, after two days, we found ourselves off the DEW Line site at Barter Island, close to the Alaska/Yukon border. A few more hours steaming and we were anchored safely in the cove at Herschel Island, being greeted by a welcoming group of people on the shore. The local RCMP detachment was certainly pleased to see us, as we were to see them.

There were two constables, as on my previous trip there on *Camsell*, plus a native constable and his wife and children. There were also about forty Siberian Huskies, of friendly disposition, undergoing training before being sent off to other RCMP posts in the North. In no time at all, we had traded some of our duty-free liquor for some of their wonderful Arctic char and some more prosaic caribou meat. We exchanged news and in turn received the latest ice and weather information. We had also noted the wife of the native constable, who was a real stunner! She was a mixture of Northern Indian and Eskimo, and exuded sexuality from every pore. We named her, among ourselves, Her Royal Majesty the Queen of the North.

Figure 45 - CSS *Richardson* sheltering in the ice pack, Beaufort Sea 1962

I invited everyone to a dinner party on board *Richardson*, with all hands involved in producing a half-decent meal. We were all agog to see more of the Queen of the North, and she did not disappoint! She was only dressed in a very ordinary blouse and blue jeans, but every curve was displayed or insinuated to full advantage. She knew how to play to an audience of men. She then proceeded to drink most of a bottle of Scotch whisky that I had put on the table, and had to be taken ashore by her sober husband. We now were aware that she was trouble with a capital T, but probably still lusted for her in our sex-starved minds. It was definitely time to move on.

We departed Herschel Island the following day for Tuktoyaktuk at the mouth of the Mackenzie River, arriving there in the early afternoon. Everyone in the village appeared pleased to see us, and after visiting the HBC wharf for mail and supplies, we tied up to the NTCL wharf close to the DEW Line site to catch our breath and prepare to set out on some real hydrographic survey work. One of the first things we had to get used to were the numerous mosquitoes and black flies, although we soon discovered that the village of Tuktoyaktuk, being situated on the shoreline and therefore a recipient of cooling breezes off the ocean, was the ideal place to get away from the miserable creatures. The village also provided an interesting view of Inuit activities, fishing, seal hunting and, by far the most dramatic, the hunt for beluga whales found in quantity close to the Mackenzie River delta. A bloodthirsty business, but a necessary part of the local culture.

Figure 46 - Ron Longbottom deploying the hand lead, shallow waters off the Mackenzie Delta, 1962

Our first task was to extend control eastward, from Kugmallit Bay along the shoreline toward the entrance to Liverpool Bay. It had been a relatively easy job when using *Camsell's* helicopters in 1960, but now we were without choppers and had to land our workboat on a shifting sandy beach and march a considerable distance inland to our chosen station. The location of the stations tended to be on a relatively high point of land thrusting up from the surrounding tundra. These high points were pingoes, of which there were many in the vicinity of Kugmallit Bay. Pingoes are earth-covered frost boils, thrust up by pressure from below acting on the permafrost. They were useful markers, but always located some distance inland and protected by myriads of mosquitoes and black flies. Our team wore protective clothing and nets, and smeared themselves with anti-mosquito goo, but never managed to achieve complete protection. We would march back out to the beach, each with our personal cloud of attacking warriors, and march straight into the ocean to submerge and eliminate the last of the persistently annoying bugs.

In the course of our travels, we rendezvoused briefly with *Camsell* and ships of the NTCL fleet in the approaches to Tuktoyaktuk, to exchange ice and navigational information. We were looked upon as an oddity, with our size and startling white hull and upper works. But they were glad that there was now a Canadian hydrographic presence in the Western Arctic.

There were a lot of caribou around, and Al Ages found himself facing a herd thundering toward him, only to veer off when he waved frantically and shouted every Dutch oath he had at his command. There was a small intermediate DEW Line site east of Kugmallit Bay with an airstrip on a sand spit jutting to sea that provided *Richardson* with some protection in northeasterly weather. This became our temporary base as we extended our work ever eastward, and commenced limited reconnaissance soundings off shore. Not entirely to our surprise, we immediately found indications of underwater pingoes at various depths.

I also had a curious confrontation with a caribou when conducting horizontal control operations in the vicinity of the radar reflector at Toker Point. While in voice communication with another station, I suddenly spotted a caribou creeping up on me through the sand dunes. I waved at the beast and shouted to no avail. It pawed the ground and snorted—I was obviously viewed as very much the intruder. The occupier of the other station could clearly hear my no doubt excited commentary and probably wished that he could record the event for posterity. After some time, the caribou disappeared in the dunes and then made a sudden charge toward me, with his antlers waving, which almost caused me to evacuate the station. Finally, I was able to complete my observations and retreat to the beach to await pickup. Even there he would not leave me alone, charging up and down the beach in a fury! I never did discover what the display of caribou anger was all about, but I was mighty thankful to be picked up off the beach in one piece.

We once found ourselves involved in a salvage operation when a large rubber tank full of fuel was spotted drifting off Liverpool Bay. It had been blown off the airstrip at the intermediate DEW Line site by a northeasterly gale, and was being sought by NTCL and the DEW Line operators. We took it in tow, with some difficulty, Al Ages acting as outrider on the dancing, undulating surface of the large tank. Eventually, with help from NTCL, we got the monster back to its home base. A very, very long day!

We also explored Liverpool Bay and the Baillie Islands, where we found signs of a settlement long abandoned. The north side of the islands was composed mainly of twenty-foot-high mud cliffs that were gradually crumbling under the impact of weather and sea ice. A graveyard and the remains of buildings were falling into the sea. There was a small, protected harbour south of the entrance between the Baillie Islands and the mainland coast. Liverpool Bay itself was quite shallow and provided little protection from the weather. By now, the weather was rapidly changing toward winter and much of our time was spent in sheltered harbours, awaiting breaks in the weather. Additionally, consistent northerly winds would drive up the water levels by as much as seven feet above normal, flooding out the land close to the shore and making life difficult for all concerned. It was during one of these enforced stays in Tuktoyaktuk that the Queen of the North briefly appeared again. She was sitting up in the bow of a skiff under the control of her husband, looking at a distance as ravishing as ever. However, on closer inspection she was sporting two lovely black eyes—her transgressions had obviously been noted!

Ages departed early to return to university, leaving us a little shorthanded. We struggled on to mid-September, with snow flurries turning to real snow falls. It was time to prepare *Richardson* for the winter. Fuel lines and water lines were drained, the wheelhouse boarded up, and the workboat made secure. All small electronic equipment and other valuable pieces were taken to the NTCL warehouse for safekeeping. We were just about ready to leave. However, we had received considerable help from Father LeMaire, the local Catholic priest, and some of his parishioners, so we invited them to a farewell meal and the remains of our duty-free liquor.

We flew out to Inuvik by float plane to await the regular PWA flight south to Edmonton. Inuvik was very much a government town, being constructed to provide the administrative headquarters for the northern Mackenzie River and the high Yukon. It had a fine airstrip, a couple of churches, a large school, and a rough-and-ready hotel, all connected to a large heated pipeline carrying fresh water lines, sewage disposal lines, and electric power lines supported on metal frames above the permafrost. It was all part of Prime Minister John Diefenbaker's vision of the future for the Canadian North.

Our flight to Edmonton took us through Yellowknife before depositing us at the Municipal Airport. The following day, we flew out by regular Trans-Canada Airline to Victoria, which was still basking in a warm fall afternoon. It was good to be home.

There were reports to write, deficiencies to be rectified, and data to be processed before sending it on to Ottawa. Among the deficiencies I noted were the Sperry gyrocompass which was located in the wheelhouse, much too high above the keel and thereby going adrift in punishing pounding weather. We also needed a radar positioning system that would enable us to use the coastal radar reflector towers erected along some stretches of the Arctic coastline. Such a system would enable *Richardson* to extend sounding coverage beyond the capabilities of line of sight.

However, I first had to make my annual debriefing trip to Ottawa. This was done, and I returned home to a happy Christmas with my family.

Chapter Nineteen

1963 – A Good Ice Year

The usual planning meetings took place prior to the season commencing. An NTCL meeting in Edmonton to start things off, followed by a planning and briefing meeting at CHS headquarters in Ottawa. Arrangements were made to have a Decca radar technician join *Richardson* in Tuk (Tuktoyaktuk) to install and test the radar reflector measuring system I visualized in 1962, in discussions with Decca radar management.

There were a number of crew changes. A hydrographic surveyor with a civil engineering background, Ron Card, would join me as mate. A proper cook was hired and two new hands joined Longbottom in the forecastle. Napier remained as our stalwart and absolutely essential marine engineer.

We arrived in Tuk by air via Edmonton and Inuvik in early June 1963. The harbour was still largely frozen over, but with melting pools of water forming over the ice surface. It was our first experience of awakening a ship from its long winter slumber in Arctic waters. Everything went well once we were able to get fresh water on board, and soon we were no longer reliant on shore-based assistance. The galley was in operation, and heat was restored throughout. Our preparations for sea went forward rapidly, while the ice surface in the harbour became ever more treacherous. I went through the ice on several occasions, before finally calling halt to the dangerous game. In early July, we were able to conduct a trial run out of the harbour. All appeared well; we were ready for another Arctic season.

Ice reports were excellent, so we undertook a voyage of exploration beyond the Beaufort Sea. Initially we went eastward to the Baillie Islands and the Smoking Hills in Franklin Bay, and then on to the vicinity of Cape Parry and the Pin Main DEW Line site, collecting hydrographic data at every opportunity. From there, we crossed over Amundsen Gulf to the Prince of Wales Strait between Banks Island and Victoria Island, exploring Walker Bay and other harbours of refuge on Victoria Island before ending up at Sachs Harbour on the southwest point of

Banks Island. It was grand fun and most productive, but extremely tiring as we were under way twenty hours or so every day. We also found that our fresh water evaporator could not keep up with demand, and that meant we were constantly on the lookout for good sources of fresh water. There were several excellent streams flowing into Walker Bay and the other bays fronting Prince of Wales Strait that met our immediate needs. We were also surprised to find some Inuit hunters in Minto Inlet, who were just as surprised to find us poking around in their territory. They were from a small settlement on Holman Island, at the entrance to Prince Albert Sound. We exchanged fresh groceries for freshly caught fish and gossip and went merrily on our separate ways. Our probe up into Prince of Wales Strait found little ice until we reached the north end of the passage, where heavy concentrations of arctic pack ice were found.

We anchored overnight in De Salis Bay on Banks Island before heading for Sachs Harbour at the southwesterly point of Banks Island

There was a small settlement at Sachs Harbour, but no chart of the harbour or approaches, so an attempt was made to carry out some preliminary survey work. However, before it could be conducted properly, a heavy concentration of sea ice drifted in from the Beaufort Sea and forced us to withdraw. This ice was the first we had encountered since departing on our lengthy voyage from Tuk. We were able to return to Tuk without seeing any more ice—it was obviously a good ice year.

Our return to Tuk was not only to refuel, resupply, and recuperate, but also to prepare to meet the Dominion Hydrographer, Norman Gray, who was traveling up the Mackenzie River on board an NTCL tug en route to Tuk. We lay at anchor, cleaned ship, and awaited the arrival of our leader. I took the chance to rest a bit as I had exerted myself a great deal during the previous three weeks. Very long twenty-hour days and the combination of master and hydrographer responsibilities on a small ship had temporarily caught up with me.

The Dominion Hydrographer arrived on board full of vim and vigour and determined to see *Richardson* in action. Our short sojourn at anchor seemed to puzzle him, almost as if we were goofing off. So away we went to conduct sounding operations east of Kugmallit Bay and to display the effectiveness of our positioning by radar while using the radar navigation beacons already established along the low-lying coastline between Warren Point and the entrance to Liverpool Bay. I believe that he enjoyed himself and was pleased with our effectiveness. Most of the remaining short season was concentrated in the Beaufort Sea east of Tuk, where our sounding efforts were delineating a number of underwater pingoes that could clearly be associated with those fully visible on shore.

Figure 47 - Sounding operations in the Beaufort Sea, author and Ron Card, 1963

Additional short voyages were made to Herschel Island in the west and to the Baillie Islands in the east. It was now early September, and the signs of winter were upon us. A series of northerly gales was sweeping the delta, and snow flurries were an almost daily occurrence. By mid-September, we were ready for winter lay-up, and soon were on our way home.

There was much data to process before my usual debriefing visit to Ottawa in December. It was also time to get to know my family again. My oldest son, Andrew, was almost eight years old and probably regarded me as a winter intruder. My wife was preoccupied with looking after our brood of four little ones. My personal ambition to succeed was being satisfied, but at a price. I had to get my life back into balance.

Chapter Twenty

1964 – A Very Bad Ice Year – Family Transported to Ottawa

Preparations for the 1964 season in the Arctic went apace, with the usual briefing visits to Edmonton and to Ottawa. Shortly afterward, my brother paid us a fleeting visit while on his way home from Malaysia to Scotland on leave. He was progressing very well in his chosen profession of planter, and had many tales to tell of events that he took part in during the Emergency in Malaya and its aftermath when the locals achieved their independence from the colonial power. He described being on an estate, miles from anywhere, where he and his workers had to be protected at night by a company of Ghurkhas, while the terrorists swept the bungalow and surrounding buildings with concentrated machine gun fire. It was no wonder he needed rest and recreation in Penang after that event. Nevertheless, his life did sound very glamorous compared to my own. When you can talk about the Sultan of Johor and the British High Commissioner on intimate terms, you have really arrived.

Gordon had a fine time while staying with us in Victoria, and he was the toast of the neighbourhood in which we lived. Before he departed, he insisted that we put on a big party with himself as host, but leaving Doreen and me to sort out the details. At his insistence, we served champagne cocktails to start off the affair with a bang—the ladies on our street were charmed and apparently still remember him with fondness even today.

After his departure, it was back to preparing for another *Richardson* year. We had a few crew changes, with Al Ages returning as mate after receiving his civil engineering degree from the University of British Columbia. Another cook was hired, and a new marine engineer. Two seamen would be hired locally in Tuk.

We had received notice that there would probably be heavy ice lying close inshore along the Arctic shoreline of the Beaufort Sea, and our first sight of the Tuk area from the air certainly confirmed that speculation. Anyhow, we had several weeks of work to return *Richardson* back to a working vessel, and that was our first priority. Our second priority was the fitting out of a number of radar reflector beacons that would interface with the Decca 404 radar in use on *Richardson*. This task was carried out personally by Lionel Dennett, the President of Decca Canada, who was intensely interested in the success of the project.

Our experiences in 1962 and 1963 had shown that the low-lying, featureless land east and west of Kugmallit Bay inhibited fixing of the ship's position by visual means to about six miles offshore. Numerous shoals, still poorly delineated on existing charts, were located well beyond this range, clearly indicating that some sort of electronic positioning system had to be employed, so that the sounding coverage could be extended further offshore and the main shipping routes adequately examined and charted.

Hi-Fix, which was effective to a range of a hundred miles, would have been ideal, but the limited accommodation and space on board *Richardson*, as well as cost and the severe problems of maintenance and supply, ruled against the use of the system or any other similar electronic positioning system.

It was therefore decided to use the radar reflector beacon system, untested and unproven in such an environment.

When *Richardson* was again operational and free of the harbour ice, NTCL made one of its helicopters available for ice reconnaissance work. The view to seaward was not too encouraging, with shore-fast ice in the bay and beyond to both east and west. It was now early July and the mosquitoes were biting ashore, but none of the NTCL fleet could proceed through the pack ice to conduct their resupply missions along the coast. My helicopter pilot fought off his boredom by chasing herds of caribou across the tundra, while I sat petrified beside him.

In mid-July we were finally able to run eastward along the coast, inside the grounded ice, to Liverpool Bay and the Baillie Islands. Everything to seaward was solid pack ice. It was not very promising. However, the ice gradually receded northward, and we were able to commence our radar/radar reflector-controlled sounding along the coast. Whenever possible during sounding operations, the radar transponder fixes were checked against sextant fixes using visual horizontal control. These checks were fairly good as long as the beacons were located within the twelve-mile range, with plotted differences in position of about one hundred and fifty feet on average. However, when one beacon was more than twelve miles distant, the range had to be determined by using the 6- to 48-mile selector, which limited the discrimination between sextant and radar fix to about three hundred feet.

Figure 48 - In heavy ice concentration, 1964

The method of fixing position by radar transponder beacons, while not as accurate as conventional means, was nevertheless sufficiently accurate for a 1:75,000 scale survey. At this scale, the space occupied by a printed sounding is several times larger than the largest difference in position between the transponder fix and the sextant fix.

Securing beacons on radar towers was found not to be good practice. The separation of four feet from the metal tower was insufficient to offset the proximity of a comparatively large mass of metal. It was also discovered that weather, atmospheric conditions, and the presence of ice appeared to affect the radar range of our beacons. The presence of ice in the path of the radar microwaves appeared to strongly affect the strength of the signals, with interruptions frequently occurring in reception when the pack ice lay along the horizon between the ship and the transponder beacons. This was probably caused by a "ducting" effect due to temperature differences over the ice. Fadeouts in signal, due to refraction or Fresnel effect, were also noted from time to time.

The ice situation continued to inhibit our plans. The launch *Quail* had been left behind in Tuk by *Camsell* to enable us to survey the approaches to Tuk from the mouth of the Mackenzie River. A reconnaissance trip on an NTCL tug appeared to prove that a deeper water channel might be found, so the launch was deployed in Kittigazuit Bay for a few days, without great success. Visibility was poor and the wind rose strongly in the afternoons, making the already very shallow waters extremely hazardous. It was decided to abandon the effort and await the arrival of *Camsell* off Tuk, where *Quail* would be taken on board for the voyage eastward to Cambridge Bay and beyond. At that point in time, *Camsell* was still stuck in heavy ice between Point Barrow and Barter Island.

Willie Rapatz had joined us in Tuk for these activities but was shortly posted south to take up duty on the *Willie J*. Al Ages was keen to take *Quail* westward to meet *Camsell* but was eventually dissuaded by the many hazards that would be encountered along the route. Almost on cue at this time, Bob Young, my boss, arrived in Tuk on his inspection visit to the region. My reservations about the ice notwithstanding, he was keen to go to sea and see for himself. We trundled eastward to the vicinity of the Baillie Islands, where we did penetrate the pack ice for a distance of twenty miles or so before running out of any semblance of open water. We could see the Smoking Hills clearly south of Cape Bathurst, crowned with a faint glow against the darker sky. These smoking hills had been burning and smoldering for centuries, a classic case of spontaneous combustion. Then the fog swept in, and our radar became our only means of navigation and that, tentative at best, with odd signals bouncing off lumps of ice and indicating open water where there was none. *Richardson* worked her way slowly back to the entrance to Liverpool Bay and anchored in the clear water between the Baillie Islands and Cape Bathurst looking forward to a restful night away from the arctic pack.

Alas, it was not to be. In the early hours of the morning, I was awakened by the tinkle of ice on our anchor cable, and found to my horror that it was the precursor of many large ice floes barreling down the passage between the mainland and the Baillie Islands. Our anchorage was not secure—we had to move up into Liverpool Bay, where shallower warmer water prevailed. There was nothing to do but resume sounding operations east of Kugmallit Bay and show Bob Young that we were not sitting on our duffs while away from home. I think that he enjoyed himself, even if it had been frustrating—we polished off his bottle of whisky with great gusto.

Returning to the deployment saga of the radar transponder beacon, it is worth noting that several improvements in technique were innovated in 1965.

Another attempt to get beyond Cape Bathurst was undertaken in late August. With great difficulty, in ice and fog, we penetrated Franklin Bay and worked our way up to Cape Parry, where we took temporary refuge in the small Cape Parry harbour south of the DEW Line site. Our reconnaissance soundings around this area were later to prove useful when Cape Parry and vicinity became the base of operations for oil and gas explorations in the Beaufort Sea.

At the termination of the season, I was asked to travel to Coppermine to assess the nautical charting requirements. I had to fly by land-based plane from Inuvik to Yellowknife, where I was supposed to take a float plane into Coppermine. It was late September, and the harbour at Coppermine was no longer open to float planes, so I thankfully was able to head south to Edmonton and home. If I had managed to get into Coppermine, I would have been delayed for some time while aircraft changed over from floats to skis. It was a silly instruction and should never have been issued.

When I reached Victoria and my family, I was faced with a new challenge. I had been selected to take part in the Senior Survey Officers Course organized by the departmental Surveys and Mapping Branch in Ottawa. The course commenced in late October, so I had to commit almost immediately. All my expenses would be paid, but if I took my wife and children with me, it would be at my own expense. It would be a costly venture, but we needed to be together as a family, so arrangements were made to rent our house and travel by train to Ottawa. It was a major expedition moving a very pregnant wife, four small children, and myself, on board a CPR train in Vancouver, to travel in cramped quarters all the way across the country to the nation's capital. At the end of three days, we were as glad to leave the train as, I am sure, the train conductor and his staff were glad to see the last of us.

We had arranged accommodation in a triplex located just outside the Ottawa city boundary. It was sufficient for our needs but hardly luxurious, and affected by the heavy truck traffic passing by. We were located on the middle level and could hear all the movements of the family above and the exuberance of a small boy below as he raced his toy car back and forth across the cement floors of the bottom apartment. Hopefully, our clan drowned out all these competing noises and evened things up.

I reported into the Surveys and Mapping classroom that was to be the focal point of my attention for the next five months. There were fourteen individuals in the class, from the Geodetic Service, Topographic Survey, Geological Survey, Legal Surveys, Army Survey Establishment and the Canadian Hydrographic Service, all opinionated and mostly anxious to learn. A few were older and perhaps there for political reasons, but most were relatively young, keen, and ambitious. There were four of us from CHS—Mike Eaton and Larry Murdock from Atlantic Region, Barry MacDonald from Central Region, and myself from beyond the Rocky Mountains.

The group mostly jelled together quite well and soon settled down into a mode where we were learning as much from one another as from our instructors. These instructors came from the various divisions of the branch, and concentrated on refreshing and rekindling our knowledge of geodetic theory, mathematical juggling, photogrammetric principles, and hydrographic and oceanographic studies. In addition, we were exposed to games of chance theory and a host of other exotic subjects. I found the interaction between us, the CHS surveyors, to be the most productive of all. Whether we fully realized it or not, we were engaged in examining our service very closely and finding it in part wanting. Eventually, we would get around to thinking about possible ways in which we could influence change in CHS.

Outside of our own group, two men contributed greatly to the success of the course. One was Hans Klinkenberg of the Geodetic Service, who lectured on geodesy and mathematics with great enthusiasm and verve, and made one feel that knowledge was a precious thing. The other was Lou Sebert from the Army Survey Establishment, who was a course participant, but with a different slant on many matters under discussion, and capable of turning a subject on its head if he felt it deserved such treatment. I am forever grateful that I knew these men.

Suddenly, Christmas was upon us, and my attention returned toward family. Friends in the Ottawa area helped make the holiday season a pleasant one, particularly for the children, who were still not sure where they fitted. It was also a trying time for my wife, who was in the final stages of her pregnancy and most anxious to get the entire business successfully concluded.

In the early hours of the morning of the December 27, I was awakened by my better half, who was ready, indeed more than ready, to head for the maternity ward. I dropped her off and was no sooner back home when she came on the phone to advise me that we had a second daughter—Sarah—who was healthy and yelling her head off. I awoke the other children to give them the good news.

Mother and daughter were back home in early January, and I could return to the course. I should, however, note that particular day of her return, because my wife was a sight to behold when I arrived to drive her and the baby home. She had combined her pregnancy with an operation to strip off her varicose veins, and a smaller one to remove a cyst on her wrist. As a result, when I arrived with our children and the neighbours' children in the car, and she appeared in a wheelchair swaddled in bandages and carrying her precious bundle, the Commissionaire who assisted her to the car murmured to himself "God bless you, Mum, and all your breed," while staring at me in accusation!

Figure 49 - My gang! 1965 – baby Sarah at home

Chapter Twenty-One

Exposure to the Wave of the Future – A Blow to the CHS Ego

In the early months of 1965, the Senior Survey Officers Course exposed us to a number of advances in technique and technology taking place in Canada and the United States. In Canada, the National Research Council was engaged in the development of many advances in photogrammetry, while the Canadian Hydrographic Service was contemplating fitting one of their larger vessels with automatic plotting tables and ancillary computer systems to control the plotting tables. All very exciting, and a continuing spur to our concern that we were fast entering a new age and that we were not ready for it.

In early March, we all received an award of sorts, when we were flown down to Washington, D.C., on a government aircraft and exposed for a week to all the new technology and techniques under development at various US Government agencies. The weather was grand in Washington, comfortable temperatures and early flowering shrubs, a welcome change from cold and bleak Ottawa.

We were taken to the US Coast & Geodetic Survey in Rockville, MD, to see all their advances in photogrammetry, electronic distance measuring, and advances in ship design relating to computer-driven plotting tables and the construction of a moon pool on board to enhance diving operations. Then we visited the US Naval Hydrographic & Oceanographic Office for further exposure to their latest charting activities. There we saw obvious signs that the US Navy was preparing for an expansion of activities along the coastlines of Viet Nam and Cambodia. Additionally, a new on-demand colour printer had been developed for field operations.

Other visits were to the US Geological Survey, the National Oceanic and Aeronautical Administration, the US Army Survey Establishment, and other linked laboratories. We were briefed on many matters of international importance and treated as respected allies of the United States.

We took part in a briefing on activities in our own Canadian Arctic, where the paths of US Navy submarines were clearly shown moving through our waters. On another occasion we had a demonstration of the high-flying U2's camera and its ability to delineate objects on the ground. Finally, we had a chance to see some experimental measuring work being conducted by laser and were duly impressed.

I cannot speak for all the others, but I do know that Mike Eaton and I were quite enthused by what we saw. There would be much to discuss in the future. My remaining time in Ottawa was usefully spent talking to people who felt like me about preparing for the future, Adam Kerr being the keenest voice.

While still in Ottawa I authored a technical paper on "Submarine Pipelines in the Western Arctic" depicting possible pipeline routes across Amundsen Gulf and Dolphin & Union Strait, with depth profiles obtained from *Camsell* and *Richardson*, together with appropriate ice and weather data. The report was made available to the oil industry, which was interested in exploration for oil and gas in the Arctic islands.

We flew home to the West Coast, none of us being up to another long journey by rail. I then set out to hand over Western Arctic operations to Al Ages, it being determined that I was due a lengthy spell of home leave. We took the children to Vancouver and Seattle and then had a longer holiday at Qualicum Beach, a pleasant spot north of Nanaimo on Vancouver Island.

I assisted on the *Willie J.* when she conducted surveys off Nanaimo later that summer. Ralph Wills was hydrographer-in-charge, and George Billard was Master. As it was my rotation year, I expressed an interest in visiting the US Coast & Geodetic Survey regional office in Seattle, to evaluate their progress in automated collection and processing of nautical data, and to establish liaison at the local level. This was achieved in October 1965, when I had the opportunity to visit. I had a tour of their plotting centre and also boarded a number of their vessels, such as *Surveyor* and *Pathfinder*, together with several smaller craft. It was a worthwhile visit, which led to a greater exchange of information between our Victoria office and Seattle, but which also led to a useful exchange of technical personnel over the years. At that time the US approach to data logging and processing was certainly in advance of anything contemplated in our Victoria office, but was not too different from what I had been exposed to at our CHS headquarters in Ottawa. A full report with recommendations was prepared for Victoria and for Ottawa. As self-appointed R&D man for the region, I also visited the US Army Corps of Engineers in Portland, Oregon, and many private sector companies engaged in developmental work relating to hydrography in both Washington State and in the Vancouver area. This information was to prove useful when a permanent R&D officer was appointed to the region a couple of years later.

Figure 50 - Surveying from the upper bridge of the *Wm. J. Stewart*, 1966

Additionally, I became a member of the Rotary Club of Douglas in Victoria, having been introduced to the club by my brother-in-law, Charlie Lowe. It was an experience that I enjoyed—mixing with a broader community of people—and becoming involved with helping others less fortunate than me.

There was also time to complete a detailed report on "The Use of Radar Transponder Beacons for Hydrographic Survey Control," based on our *Richardson* Arctic experience of 1964.

In the autumn of 1965, CHS created three new senior management positions—assistant regional hydrographers—in all three regions, Atlantic, Central, and Pacific, and invited a number of us to apply for the positions. Eight senior survey officers were selected to go before a board set up in Ottawa: DeGrasse, Corkum, Eaton, Blandford, Kerr, Wills, Sandilands, and me. The board was chaired by Dr. W. Cameron, director of the Marine Sciences Branch. He was assisted by Norman Gray, Dominion Hydrographer, Clarence Cross, head of Tides and Water Levels, and others. Bill Cameron made it clear at once that he was in charge. He reminded the board and later each trembling supplicant who came before the board that he strongly endorsed the view that the hydrographer's charting tasks should be extended to describe other parameters of the sea bottom, and he suggested that the hydrographer might eventually move to describing the sea itself as it varies from place to place and time to time. He then orchestrated the board in its utter demolition of the eight candidates who had dared to present themselves as being worthy of consideration for these new management positions. It was "blood on the carpet" as one member of the board later described the event. I called it "Slaughter on Booth Street." The eight most qualified members of CHS had been told in no uncertain manner that we all lacked the training and education to meet Bill Cameron's exacting standards. What a slap in the face, not just personally but to the entire CHS.

Those of us from the west coast and others retired to a suitable drinking establishment to lick our wounds and ponder the future. We were angry and despondent, but determined in some way to fight back. Some sort of a joint response was required, but it would take time to formalize and liaise with CHS management. I thought about discussions that I had had with other CHS officers after returning from our Washington sojourn, and retained some immediate comfort. It so happened that Ian Miller, an ex-hydrographer, was hoisting a few with us, so we got the benefit of his sage advice: "Keep pushing CHS management; you will eventually get what you are seeking." He was an interesting chap, a former mariner who had become an expert in Decca positioning systems and was now employed by a US Avionics company. He was a seafaring Scot who had survived the War and then taken command of the Royal Yacht *Baghdad* in Basra—the personal toy of King Faisal of Iraq. He later published a very interesting story of his life's adventures.

We retired to never-never land on the west coast to sooth our battered egos while surrounded by the solace of our respective families. I soon, however, was having meetings locally with other interested parties regarding the need for increased training and education within the CHS. I was also in frequent telephone communication with Mike Eaton and others across the country. It was the genesis of our Canadian Hydrographers Association, later to become the Canadian Hydrographic Association or CHA.

All of the foregoing was bad enough, but on top of the concern about the "uneducated hydrographer," we were faced with a Bureau of Classification Revision exercise to slot each and every surveyor into either a survey officer classification or a survey technician position. A slogan was being pushed—"Too many Chiefs and too few Indians"—to justify the massive undertaking. According to this philosophy, the Indians would be the button pushers of the future and the Chiefs would be the architects of modern hydrographic surveys, expanded in scope. All this was very fascinating, but that future was a long way from being achieved by this clumsy exercise. People frantically wrote and rewrote and rewrote their BCR documents to ensure that they were accepted as being a Chief or, at the very least, a high-ranking Indian. What games we play! And all of this brought about by the Glassco Commission on the Civil Service, which actually proposed "Let the Manager manage," but saw its proposal perverted by the bureaucrats in the Treasury Board and the Civil Service Commission. To the "massacred" and others in CHS, the BCR exercise and the need for additional training and education appeared to complement one another, but also appeared to make the end objective more difficult to achieve.

Our informal meetings and discussions in 1966 within the CHA entity as proposed, and with members of the CHS management team who had some sympathy for our aims and objectives, led to a discussion paper drafted by Mike Eaton that was widely circulated nationally among interested hydrographers. He expressed his own views on aims, knowledge required, attitudes affecting education, methods to achieve stated goals, and finally suggested a programme that the Marine Sciences Branch might adopt. All good stuff—but requiring a strong management commitment.

During this period, the "massacred" were being encouraged to upgrade their education in mathematics and physics. Instead of being shipped back to sea, many of us found ourselves sitting in stuffy university classrooms, pondering problems that were probably a breeze to our very much younger classmates.

Just in case anyone thinks that my sole activity in 1966 was my preoccupation with the training and education of the hydrographer, I am proud to acknowledge that I managed to produce a paper on "Navigation in the Western Arctic," which was an update on navigational information and advice in the area compiled from on-the-scene notations recorded while on *Storis* (1958–59), *Camsell* (1960), *Spalding* (1961), and *Richardson* (1962–65).

In March 1967, at the CHS conference in Dartmouth, Nova Scotia, CHA was formally presented to CHS management, and Mike Eaton was elected as president. I presented a paper entitled "The Educated Hydrographer," which incorporated my Pacific Coast experience and views on the immediate need for a CHS-supported long-term training and education programme. The proposal was discussed in depth with CHS management and representatives of CHA, and I was instructed to prepare a detailed education brief for further review by the CHS management. I was given the assistance of a Senior Survey Training Officer, but felt that I could achieve much more working on my own.

Richardson had returned to Victoria in the late summer of 1966 to undergo a major refit. I now found myself taking over command of *Richardson* while trying to find time to develop the education brief. The brief was completed in late May 1967, for review by the Dominion hydrographer and his CHS management team, and hopefully action.

The following steps were proposed. That:

1) Refresher courses in physics and mathematics be made mandatory.

2) A junior hydrographic technical course be established, to follow on from the present basic entry course, with a three-year field service set as prerequisite. Course subjects would include survey general, sounding, projections, radio aids, navigation, seamanship, tides and tidal theory, and cartography.

3) Refresher courses—in-house or at university—be made available for surveyors with the appropriate educational background. Subjects such as physics theory on light and heat, electronics, magnetism, and marine geology would be emphasized.

4) A new hydrographic course aimed at hydrographers with HIC potential become a high priority. Subjects to include projections, radio aids—principles, errors, calibration, etc., photogrammetry, applied electronics, astronomy, computers. As a pre-requisite, five years of field service would be necessary.

5) A specialist senior hydrographer course be developed to cover the subjects of geophysics, geology, the principles of tidal analysis and prediction, and physical oceanography.

6) Wherever possible, the training be conducted by the CHA, but that in some areas university-level instruction would be required.

The foregoing steps were presented as a proposal to deal with the emergency situation that existed at that time (1967) and were intended to stimulate the interest of both hydrographers and management in an education and training programme geared to career development, and offering enhanced prestige to both the professional hydrographer and his employer.

Chapter Twenty-Two

1967 Voyage to the Western Arctic – An Epic Venture

Deep inside me, I am an old-fashioned Scot, rational on the surface but bothered from time to time with flashes of uneasiness about future events. It was so in 1967, when I resumed my *Richardson* appointment. I was happy to be off to sea once more, but wondered if a second voyage into the Western Arctic might perhaps be pushing my luck too far. The heavy ice concentrations encountered in 1964 in the Beaufort Sea and the Amundsen Gulf were also a clear indicator of changing weather and ice patterns. I also worried about radio communication equipment, which had not been replaced to meet my perhaps exacting standards. However, Terry Daniels of Victoria did his best with the radio equipment and installed a new Arma Brown gyrocompass below decks to ensure compass stability. Nevertheless, we undertook a trial voyage around Vancouver Island after completing refit in the shipyard. As noted before, I was also composing and completing my "Education Brief," so every day was a full one.

Our new crew consisted of Tony Mortimer, hydrographer and mate, Dick Lofgrin, engineer, Murray Petrie, cook, Vic Erb, coxswain, and Dave Cox and Alan Meadows, seamen. The two seamen were relatively inexperienced, but the engineer, cook, and coxswain had all been with the ship in the previous year, and so presumably had some experience. Tony Mortimer was a Master Mariner and keen to take part in our voyage. He would prove to be a tower of strength when we needed him.

Our circumnavigation of Vancouver Island was similar to that of 1962, except that we visited more ports and harbours seeking sailing directions information. On the inside passage, time was spent at Ladysmith and Nanaimo and, at the northern tip of the island, in Port Hardy and Port McNeil. Down the west coast, visits were paid to Friendly Cove, Tofino, Ucluelet, Port Alberni, and Bamfield before returning to Victoria to make final preparations for the long journey north.

With fond farewells from families and other interested parties, *Richardson* departed Victoria Harbour in early July, but not before entertaining the spectators with a few unrehearsed movements as our steering gear misbehaved. We had an uneventful run up the inside passage to Prince Rupert. There we refueled, watered, and took on fresh supplies of food, while making last-minute phone calls home. An unexpected pleasure for me was to be greeted by Adam Kerr from Ottawa who was visiting the area on some mission from headquarters. We did not, however, delay and were soon on our way into Alaskan waters.

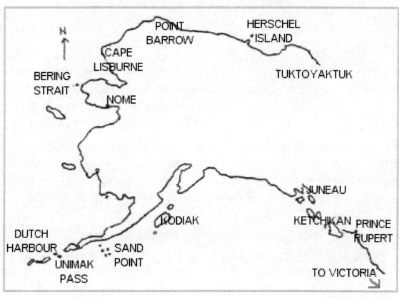

Figure 51 - The 4047 nautical miles between Victoria and Tuktoyaktuk

From Ketchikan, we proceeded northward past Wrangell and Petersburg and the ice calving off the glaciers, before arriving at Juneau, that tourist trap with a dozen honky-tonk bars stretching along the waterfront, interposed with shops selling a mixture of T-shirts and such junk, and a few more up-market ones peddling native jewelry and collectibles carved from walrus tusks. We quaffed a few ales while ogling the local talent, whatever there was of it, reprovisioned, and were soon on our way toward Cross Sound and the Gulf of Alaska. The sky was overcast as we approached Cross Sound, but cleared as the wind freshened, ensuring that our entry into the Gulf was lively. We made good time, passing numbers of fishing vessels and several very long-line operations before arriving in Kodiak harbour. On our previous visit to Kodiak we had docked in the US Naval Base, but found it to be hampered by regulation and protocol. The civilian harbour at Kodiak was quite different, being of very recent construction, its predecessor having been destroyed by a violent earthquake and its aftermath in 1964. We were welcomed by the locals and partook of their watery beer before taking in fresh water, fuel, and fresh supplies.

Our route then took us through Afognak Passage, between Kodiak Island and Afognak Island, out into Shelikof Strait and then southwestward to the Shumagin Islands, where we found refuge from a minor storm in the small island harbour of Sand Point. There we found ourselves tied up among a number of Canadian halibut fishing vessels. One of them was *Sleep Robber*, well remembered from the New Westminster Shipyard, where we shared a finger wharf.

The Captain and crew of *Sleep Robber* made us very welcome and fed us superb halibut steaks, while I produced some of our duty-free booze. We discovered that there was a small bar on the island that opened and closed for business day and night according to the arrivals and departures of fishing vessels. We paid the bar a visit and then finished up the festivities in the wee small hours on board *Sleep Robber*.

It was a sad captain and crew of *Richardson* that gingerly made its way out of harbour in the first hour of dull daylight. We ran into fresh westerly weather as we approached Unimak Pass, but ran on to Dutch Harbor, where we anchored overnight and bought a number of huge king crabs straight off the crabbers unloading their cargoes into a gigantic factory ship. Another top-up of freshwater tanks, and we were heading north up the Bering Sea toward Nome. The weather was rough until we reached the vicinity of St. Lawrence Island, and then eased as we approached Norton Sound and attempted to anchor off Nome. The swell ran high, and it was rather uncomfortable, so we headed eastward into Norton Sound and found a safe anchorage in Golovin Bay, where we had anchored and watered in 1962. We were able to poke our bow under a small waterfall and allow everyone to take showers. Our radar had not been performing well on the run up the Bering Sea, so the next order of business was to attempt to find the problem and fix it. After a full day's work, it was operating again but not at its usual efficiency. No help could be had from Nome, as we could not raise the local radio station and, in any case, we drew too much water to get over the bar into the harbour. Reluctantly, I decided to take a chance and continue northward.

Camsell had dropped off a cache of fuel in drums at Port Clarence, just south of the Bering Strait. Having taken on the fuel, we headed up through the Bering Strait, while the radar signals continued to be inadequate. We therefore took temporary refuge in Kotzebue Sound, anchoring inside Cape Esenberg for shelter. The water level was quite shallow and affected by the prevailing east-northeast winds, making it an uncomfortable and ill-protected anchorage. Nevertheless, we got out the Decca radar handbook, and attempted to find the fault and carry out the needed repairs. This exercise was carried out between shifting anchorages several times and, finally, after the best part of two days, I was reasonably convinced that the radar would see us through the ice barrier up ahead. The problem apparently was water in the wave guide reflector, detected and fixed by Tony Mortimer.

Figure 52 - CSS *Richardson* in trouble – vicinity of Point Barrow, July 1967

Richardson rounded Cape Lisburne into the Arctic Ocean in clear weather and with no signs of ice, but by Point Lay the ice edge was on the horizon. By the time Icy Point lay abeam about ten miles distant, we were into the ice. It consisted mainly of large floes about six feet thick, but with clear leads opening up as we pushed our way toward Point Barrow. Off Point Franklin, we ran out of leads and could not retreat. There was a strong southwesterly wind developing, and the entire ice pack seemed to be moving northeastward, parallel to the shore, at about 0.5 knots. Off Skull Cliff we had really become beset in the ice. The workboat was fully equipped with emergency gear and landed on the ice surface close to the ship. On July 22, the ice started to come under enormous pressure, due to a combination of winds and tidal currents, with the ice rafting and whole floes stacking themselves one on top of the other. The workboat was lifted from the fast deteriorating ice on to *Richardson*. I advised *Camsell* by radio that I might be needing his assistance.

Shortly afterward, *Richardson* herself came under enormous pressure, and we lurched from side to side as the solid ice cover disintegrated around us. All hands clung to whatever support was available, as the clinometer registered fifty- to sixty-degree heels from the vertical. Sitting on the deck of the wheelhouse, with legs wedged up against the chart table, I reluctantly sent out the Mayday call. Shortly thereafter, we were advised that *Camsell* would be coming to our rescue from the Beaufort Sea and that the USN icebreaker *Northwind* was on her way from the Chukchi Sea. Next, I was contacted by the Point Barrow DEW Line site, who wanted to know if they should order up a large military rescue helicopter from Elmedorf Air Force Base in Southern Alaska. It would take a day or so to arrive in the Point Barrow area—were we really in such desperate shape? This bargain basement approach to our rescue struck me as being hilarious, and I advised Barrow that I would consider the offer and get back to them if our situation further deteriorated. After all, I had *Camsell* and *Northwind* up my sleeve!

Actually, the situation did get worse shortly afterward, when rafting ice pushed our hard-done-by vessel up and out of the water completely, leaving us without engine power, galley power or heating. We bundled up and existed on bread and bologna sandwiches. We could now see Barrow village in the distance—about eight miles away—and became the target for a series of flybys by light aircraft and helicopters. We knew then that we must be on the news at home. As the Barrow DEW Line Site reported, we "were drifting along in the northerly flow—being bowled along by the crushing ice."

Figure 53 - Captain Strand from *Camsell* comes calling, July 1967

By the July 24, we had drifted about ten miles north of Point Barrow, still beset in heavy ice, which was tightening its grip on us. Our propeller and rudder were badly damaged, but hull damage appeared to be minimal. Camsell appeared to the eastward of our position and soon *Northwind* showed up to the southward. They fought their way to within a couple of miles of us before succumbing to the same pressure ice that held us in its grasp. In the early hours of the July25, Captain John Strand of *Camsell* landed by helicopter on the ice close to us and we drew up a plan of action that would involve all three vessels. It was good to see him and the chopper pilot Bob Masters—just like old times back in the early sixties. The pressure was coming off the ice as we all drifted northward.

Figure 54 - *Northwind to the rescue*, July 1967

We joined up in a train, with *Northwind* in the lead, followed by *Camsell* with her bow buried in *Northwind*'s stern towing slot and, finally, the caboose—little *Richardson*! We could not slot into *Camsell's* stern towing slot, as our bow was only a few feet above the waterline. So we secured ourselves as closely as possible to *Camsell*'s stern and hoped for the best. An almost constant companion during these delicate manoeuvres was Art Mountain, the First Mate of *Camsell*, who looked down on *Richardson* from the stern of *Camsell* to ensure that we were still there. With his dark parka hood adorning his tall spare frame, he reminded me of a worried but friendly bird of prey. He was there for us!

The train concept worked for a while, when *Northwind* was proceeding slowly with due caution. Later on, however, *Northwind* got the bit in her teeth and speeded up before smashing into an ice pressure ridge and coming up all standing. *Camsell* absorbed the shock with her bow arrangement, while *Richardson* smashed her bows into *Camsell's* stern and began to take real punishment to her bow and flying bridge. At this point in time, an altercation took place between *Northwind*, *Camsell*, and *Richardson*, with the captain of *Northwind* strongly suggesting that we abandon ship and *Northwind* would then use *Richardson* for gunnery practice. Meanwhile, *Camsell* advised that we abandon ship for our own safety. I was apoplectic when receiving those messages and indicated strongly that we were not abandoning ship under any circumstances. So, the train resumed its slow progress and *Richardson* continued to absorb a lot of punishment. We arrived off Elson Lagoon later in the day and, in the safety of the anchorage, undertook essential repairs with the assistance of experts from *Camsell*.

Figure 55 - The Arctic train: *Northwind*, *Camsell*, taken from little *Richardson*, July 1967

Nothing could be done about the damage sustained to our bow and rails, but *Camsell*'s divers were able to temporarily repair our battered and twisted propeller and rudder. A planning meeting was held with Captain Strand, where I hesitatingly agreed to accept his offer of a tow through the ice pack to Canadian waters. I favoured a solo performance, repeating my success of 1962 by running along the coast inside the grounded ice edge, and thereby avoiding the ice pack. However, as Captain Strand indicated, if we broke down we would be on our own as *Camsell* had to return to the Beaufort Sea to provide ice breaking escort to the

DEW Line supply ships. So off we went in tow through the leads that *Camsell* found in the heavy concentrations of ice.

Figure 56 - The long tow along the north coast of Alaska, July/August 1967

On the July 29, we stopped in a large area of open water among the ice, with thick fog prevailing, and went alongside *Camsell*. There, the divers secured the keyway cap to the rudder post and replaced our twisted propeller with the spare. Our crew were invited on board *Camsell* for showers, a hot meal, and the pleasure of seeing a movie. In the middle of the movie, I was alerted to danger when the lookout reported that the ice had moved in and closed off the open water area. The ice was pushing hard up against *Richardson*'s hull, forcing our upper works and lower rails into contact with the solid hull of *Camsell*. Heavy damage ensued and, for the first time in many days, I felt a sense of despair. However, the fog cleared, and *Camsell* was able to tow us into more open water, north and east of Barter Island. We advised *Camsell* that we would attempt to proceed on our own. *Camsell* then departed with our sincere thanks.

She had left us several drums of fuel that we attempted to transfer into our tanks, only to find that they were contaminated with water. A long hard spell of work ensued, pumping by hand and sieving the fuel through our Stetson hats before all was well once more. Then the keyway cap on the rudder post gave way, and several hours of work were spent before we got underway once more at reduced speed. In largely open water, we crossed Mackenzie Bay and rounded the delta before entering the harbour of Tuktoyaktuk. It was now August 1, 1967. The floating dry dock was prepared for our arrival and we were soon up on the stocks awaiting repairs. It had been quite a voyage!

When we received mail from home, we realized that our exploit had garnered much attention from the media, in Victoria particularly, but also in Vancouver and the broader Pacific Northwest. My wife's first notification of our troubles was a Victoria radio station report citing an Alaskan source for the news. Later, she was phoned by a reporter from a Vancouver newspaper who apparently assumed the worst and expressed his sorrow at her being left alone with our five children. She told him in no uncertain terms that her husband was not dead yet and abruptly terminated the conversation.

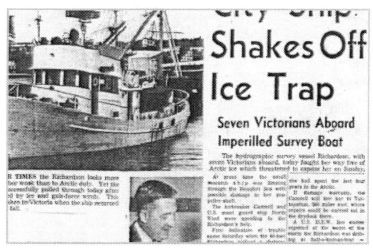

Figure 57 - The news from home, 1967

The newspaper clippings in later letters gave a blow-by-blow description of our predicament, sometimes fairly accurately, but often overblown and speculative. Some of the best articles were Arthur Mayse's column in the *Daily Colonist*, where he used material provided by Alan Meadows' mother from her son's descriptive letters to her directly from the scene of the action. These clippings were of great assistance to me in reviewing and recalling the events of that period. I had some notes, but my ship's log book vanished into a great hole in Ottawa and was soon shredded by some idiotic bureaucratic fiat. The personal recollections of those on board became all important. I am particularly indebted to Tony Mortimer and Alan Meadows for their input into the saga.

The repairs required to bring *Richardson* back to operational status were quite extensive, and so it was August 13 before we were fit to commence hydrographic operations. Our hull looked like washboard before the plating was replaced. We acquired a new propeller and returned our rudder and keyway cap to pristine condition. Additionally, the bow, the deck railings, and the flying bridge were all repaired to meet MOT specifications, while the ship's engines and auxiliary equipment were pronounced fit for sea conditions.

With such a shortened season, we concentrated our survey efforts on the Warren Point to the Baillie Islands area, using the radar reflector measuring system previously employed. The survey paid dividends when we were able to delineate the limits of shoal waters a few miles north of Atkinson Point. These were probably minor underwater pingoes, as they appeared offshore of a cluster of onshore pingoes due east of Tuktoyaktuk.

Figure 58 - A family welcome at Victoria Airport, September 1967

Another concern was support to *Beluga*, a new Bertram inboard/outboard twenty-foot launch that had arrived in Tuk via the river in damaged condition. Barry Lusk was in charge of the survey of Kittigazuit Bay, the shallow water area lying between Tuk Harbour and the mouth of the Mackenzie River. It was a frustrating task that we supported to the best of our ability. The damaged outboard leg was constantly breaking down, requiring *Beluga* to attempt to return to safe haven by stern power only. Additionally, horizontal control was difficult to establish and to see through sextant telescopes, and a thirty-knot wind whipping up the shallow waters made it an almost impossible task. It was *Richardson* to the rescue on several occasions. Barry Lusk had said how awful we all looked when *Richardson* survived her ordeal; it was now our turn to repay the compliment.

A break from normal routine took place when Charlie McIntosh arrived in Tuk for a short visit to familiarize himself with *Richardson* and her area of operation. Apparently, he was scheduled to take my place in 1968. Cox departed early to return to university and a local lad was hired in his place.

With winter upon us, we laid up *Richardson* and departed Tuk on September 21. We were safely home in Victoria on September 23. My family was in full strength at the Victoria Airport to greet me, supported by our neighbour Arch Snow and his CBC photographer. We were ensured nationwide television coverage and additional comment in the Pacific Coast press.

The only person who did not approve of the attention was my youngest daughter Sarah, who hated her daddy's full beard. We survivors of the Point Barrow episode gathered later at my home for a brief celebration.

Figure 59 - Sarah's reluctant welcome, 1967

The response of the Dominion Hydrographer to my CHA Education Brief was awaiting me, and I was instructed to proceed to Ottawa to discuss the matter in greater detail after a few days at home with my family.

Chapter Twenty-Three

Changes in Direction – Becoming a Desk Jockey

The response of the Dominion Hydrographer to the CHA Brief on Education was encouraging, but suggested that our approach be modified into three steps: Entry, Intermediate and Senior levels, of course. It also outlined a number of items that might be of assistance in implementing our proposals—tuition costs, study time, assignment of in-house lecturers, and an emphasis on the use of the rotational year for a major part of the staff educational programme.

A meeting of CHA representatives and CHS management was suggested for October seeking consensus and clarifications on the following points:

1) Recognition of the need for formal training

2) Details of syllabi

3) Standards of courses including examinations

4) Prerequisites to enter each step

5) PSC regulations governing tuition fees etc.

The CHA representatives present at the meeting were Mike Eaton, president, myself as vice president, Dick LeLievre as secretary, and Adam Kerr and Ab Rogers from Central Region. CHS Management were represented by Norman Gray, the Dominion Hydrographer, Colin Martin his deputy, Mike Bolton, Regional Hydrographer Central, and Russ Melanson, Regional Hydrographer Atlantic.

A compromise was reached on the number of steps required, as suggested by the Dominion Hydrographer. The CHA step 1 and step 2 courses would be amalgamated with the existing CHS entry-level course. The CHA step 3 (refresher course) and CHA step 4 (senior technical) course would be amalgamated to become the CHS step 2, or advanced hydrographer's course. The

CHA step 5 (university level) needed further study, and a decision was deferred for the time being. The readiness of CHA to conduct its own examinations was a matter of discussion and left in abeyance for the immediate future. The first CHS step 2 course was agreed upon in content and was scheduled to be held in the summer of 1968. The CHA explained its position and status to CHS management at their request. CHA was a professional body established to promote the professionalism of the Canadian hydrographic surveyor and was not a union fighting for salary increases or better living conditions for its members. Neither was it a social club. Such a declaration seemed to meet with the approval of CHS management.

After all the preparation and intense discussions about hydrographic education, I needed a rest, but it was not to be. Along with a few others in the surveying and mapping field, I had been selected to take part in an intense study known as "The Managerial Grid." The study would be one week long and would take place at the Holiday Inn in Peterborough, Ontario, about one hundred miles east of Toronto. My colleague Adam Kerr very kindly volunteered to drive me to our destination.

There were about forty of us on the course, divided roughly into teams of six persons. At that time in the late sixties, few women were in the managerial ranks—there was a solitary one among us. The object of the course was to ascertain where each manager fitted on a grid that depicted a score of 9 as the ultimate in managerial concern for job at hand and for the staff employed on the project at hand. A score of 1 alternatively indicated low concern for productivity or personnel. Other graduations were permitted but indicated less than full commitment. The course staff instructed and indeed dragooned each team into revealing its innermost thoughts about management and thereby brought about a hierarchy of leaders and followers. For me it became demanding and exciting, and for all of us it culminated in being rated by the other members of my team. I may have thought that I was the perfect manager, but my companions rated me as a 9/1 with a 9/9 in reserve. In other words, I was a driver with small consideration for my staff, but would revert to dual concern for the task at hand and my employees if that was what it took to achieve success. A Captain Bligh with a Captain Cook in reserve. Revealing indeed!

Just before the Managerial Grid experience terminated, I received word from my wife that Malcolm, our youngest son, was very ill but seemed to be on the road to recovery. I was therefore most anxious to return to the west coast, rather than complete my work in Ottawa. So I hastened back to Victoria to find my wife still worried but Malcolm recovering well. He had lost a lot of blood due to a botched operation on his tonsils and was perhaps lucky to be still alive. Now we were a family once more!

Returning briefly to my impressions of the Managerial Grid, I could not help but see the many and varied reactions to exposure among our team and, from gossip, among the other teams. One member of my team was most hurt at his rating by team members and seemed to go into a bit of a depression, another became most combative, while a member of another team withdrew to the balcony because he felt that his team had not assessed his managerial capabilities properly by failing to take into account the fact that he had a Ph.D. in the Sciences. Our one female participant was accused of operating behind a façade, which caused another row. Perhaps the Grid was not the best and only way to rate Managers.

Out on the coast the winter passed rapidly. In addition to my CHA/CHS Education brief, I continued the job of attempting to establish a small Pacific Region research and development presence that would mirror the activities I had already witnessed in Ottawa and at the Bedford Institute of Oceanography in Nova Scotia. I was ill trained for this sort of work, but no one else appeared interested, so Bob Young gave me a fairly free rein. I again visited the US Coast & Geodetic Survey establishment in Seattle and the US Corps of Engineers office in Portland, Oregon, seeking information on new technology and techniques that could usefully be adopted by the Canadian Hydrographic Service. Additionally, I visited private-sector companies whose own R&D work in positioning and in underwater acoustics showed promise that could be of use to CHS. My work was very much of a preliminary nature, but would be useful in the future when CHS Pacific Coast adopted a more positive approach to its own research and development needs.

I also managed to find time to complete an updated report on the *Richardson* Decca Radar Ranging Unit and the Transponder system, which included an evaluation of the system's strengths and weaknesses covering the period 1964 to 1967. It should be noted that the ARU and Beacon system continued to serve *Richardson* well for several more years, augmented by the Polar Continental Shelf Project Decca 6f chain established to cover the Beaufort Sea area in 1969.

In early March 1968, the Canadian Hydrographic Service conference was held in Ottawa. It was also the occasion of the 2nd CHA conference, which elected me as their new president. I was pleased but knew that much work remained to be done to ensure the adoption of the CHS/CHA three steps to the Educated Hydrographer. Although there were strong supporters of our proposed programme, there were doubters in the ranks who had to be persuaded of the righteousness of our cause and, in between, a larger mass of the undecided who still sat on the fence. It would have made our task in dealing with CHS management that much simpler if we could present a strong united case. In my darker moments, I felt like banging a few heads together!

Then, at the end of the CHS conference, a startling event occurred. Dr. Cameron, the Director General of the Marine Sciences, announced two new appointments. With the coming retirement of Norman Gray as Dominion Hydrographer, his place would be taken by Dr. Art Collin, and with the expected retirement of Bob Young as Regional Hydrographer Pacific, Tom McCulloch would be appointed as Regional Hydrographer designate – region to be announced later. There was a hush in the hall, followed by a babble of noise as the DG strode off the platform. I was stunned! I had felt that things were looking up for me, from appreciative words from senior officers and other intangible comments. But I never expected that announcement. In short order I received sincere congratulations from some colleagues, and saw consternation on the faces of others, with some avoiding me altogether. The price of success was already becoming apparent.

The following day I was summoned to meet with CHS management and formally asked if I would accept the position offered. It was suggested that I might wish to have a few days to consider the matter and consult with my wife. Without hesitation I stated that I accepted the position offered and so set a series of actions in motion. While the formalities were handled as expeditiously as possible, there were formal appeals against the method of my appointment, which under the circumstances was hardly surprising. It was mid-May before I could safely start to plan to move our family to Ottawa, where I would take over as Regional Hydrographer Central from Mike Bolton, who would become Regional Hydrographer Pacific.

I made several trips to Ottawa in late May and into June to familiarize myself with my new job, getting to know my staff and, in particular, Harvey Blandford, the Assistant Regional Hydrographer, who was a great help to me in those early days of my new responsibilities. He too was a Master Mariner, which meant that we shared a common view on the all-important link between the mariner (the end user) and the contents of our end product (the nautical chart). Our offices were in the City Centre Building, a short couple of miles from CHS headquarters on Booth Street. This proximity of our regional office to CHS headquarters bothered me, whenever I was drawn into decisions and had to spend a great deal of time moving between the two offices. I appreciated being given the opportunity to provide input on national matters but was much more anxious to learn more about my region, its boundaries, and the diverse techniques that were being deployed in the inland waterways and in the High Arctic operations of the Polar Continental Shelf Project.

On each visit I stayed at an apartment hotel, just off Elgin Street, close to the heart of the city. There I could look after myself in relative comfort while trying to catch up on the many files that I had to peruse and hopefully understand. Even there, I was apt to be interrupted by the arrival of keen young hydrographers who wished to press their case for action on their pet project. One such was Neil Anderson of the Polar Continental Shelf Project and an enthusiastic supporter of the CHA position on education. He was so keen that he brought along his very pregnant wife for support. The poor woman must have suffered in the almost tropical heat of an Ottawa evening, while having to listen to her husband and me ramble on about technology and technique and how a better-trained and educated hydrographer could exploit the connection. He was the region's R&D man and a great believer in in-house R&D, as opposed to selecting new equipment off the shelf and attempting to make it fit a given situation.

While all of this was going on, Ottawa was in a bit of a turmoil, with the Liberal Party about to select a new leader. My residence appeared to be full of potential candidates and their supporters. Their late-night parties and these visits I received from enthusiastic members of my staff took quite a toll, and I was glad to return home for a few more relaxing days. In late June I made my final short business trip to Ottawa, and then concentrated on moving us all lock, stock, and barrel to the nation's capital. Our house on Ten Mile Point was sold, several parties were thrown, the children said tearful farewells to their friends, and Doreen and I said our sad goodbyes to both old friends and family We had made many friends during our years in Victoria, our children had strong links to their school chums, and while Doreen had family and faith as support, I had become a Rotarian and a committed member of the Conservative Party. All these links had to be gently severed before we were ready for our great journey across the continent.

I set off first with our three boys in tow, with the car filled with travel gear, and a pact agreed upon. Each motel we stopped at for overnight rest must have a pool and good air conditioning. The pool for the kids and the air for the weary driver. It worked our fairly well, although the children did not really appreciate our very early morning starts. Our first stop was Penticton in the Okanagan valley. Then, after eating pounds of fresh cherries around Vernon, we pressed on into Banff for our overnight stay. We skirted Calgary as it was Stampede time, lunched in a very hot Medicine Hat, and found evening paradise on the outskirts of Swift Current. An early breakfast in Moose Jaw, and we were on our way through Brandon to

Winnipeg, where we found a motel that met our simple requirements. A phone call to Victoria friends holidaying at Victoria Beach on Lake Winnipeg got us an invitation to visit, so that became our next stop. It was very pleasant and relaxing, even though a few mosquitoes and black flies were present. There, however, Doreen phoned me with the sad news that her father, who had been in ill health, had passed away peacefully. Doreen's sister Joan and her family were also transiting the road to Port Arthur to visit friends and other relatives and were unaware of the sad news, so I set out from Victoria Beach to try to pass them the news when we spotted them on the highway between Rainy River and the Lakehead.

We were unable to make contact until after we had located ourselves in a motel halfway between Port Arthur and Fort William. It was a sorrowful time.

While in Port Arthur I was able to see Jimmy Graham, the former Harbour Master and a cousin of Bobbie Guerard, Doreen's sister Ursula's deceased husband. The poor chap was ailing, a condition that did not enhance my memories of the Lakehead—memories that were mostly gloomy. However, I did take the boys to many places that I remembered with pleasure, such as the Kakabeka Falls, the Lakehead waterfront, and the view from the Port Arthur Heights of Thunder Bay and the Sleeping Giant.

Our last stop before arriving in Ottawa was in Sault Ste Marie, where unfortunately we could not find a bathing pool. The weather had cooled off a bit around Lake Superior, so there were not too many complaints. Once we were past Sudbury, the weather got sticky, with frequent thunderstorms, so we were very glad to arrive at our destination, which had both pool and adequate air conditioning. We awaited the arrival of our females with great anticipation.

They arrived at the Ottawa Airport the following evening, tired and teary, with a very upset cat in tow. Poor Jezebel had traveled in the hold of the aircraft and was much affected by the noise level. We settled into a family relationship again, albeit in a motel/restaurant atmosphere, while preparing to move into an older house in Britannia Beach nearby. Our move into the house was to some extent precipitated by my discovering that one of our sons, Duncan, was ordering exotic and expensive breakfast dishes for himself and encouraging his siblings to do likewise. As a family we needed to get back to reality.

The house was fairly large, standing on a quarter-acre lot. It had been constructed solidly back in the late 1880s, with good hardwood on the floors, window frames, doors, and stairways. It had a large living room, dining room, and kitchen, together with a laundry/bathroom on the main floor. There were four bedrooms above the main floor with full bathroom, and in the attic a room that became our fifth bedroom. Jezebel disappeared into the basement, while our eldest son Andrew took possession of the attic. There was a small garden to potter around in, and lots of lawn to keep trimmed. We were ready to be part of the Ottawa scene.

In an earlier chapter I mentioned Mike Bolton, whom I first met in Ottawa in 1959. We became good friends over the years, and often engaged in discussions on what we perceived was wrong with the Canadian Hydrographic Service. These interactions were most useful when we talked about the need for a proper career training programme for hydrographic personnel. Mike was just as vociferous as I was, but he was closer to management (indeed he was management latterly), so his approach to the training situation was more pragmatic and muted than my own, and eventually probably more productive.

When we were both part of the CHS management team, we continued our friendship, sometimes enlivened by fierce argument about matters of great concern to both of us. When later I became Director General and no longer had direct input into CHS, he was still always available for discussion and opinion. I valued his comments and continuing friendship highly. He was, indisputably, my mentor in matters concerning the CHS.

Chapter Twenty-Four

Regional Responsibilities

Settled in Ottawa at last, I was about to take a hands-on look at my little empire that stretched from the lower St. Lawrence River to the Prairies in the west and northward into the High Arctic. Over the course of several months I visited CHS field operations on the St. Lawrence River in the vicinity of Batiscan, which lies about halfway between Quebec City and Trois Rivières. Our survey team operated twenty-one-foot motor launches. They were largely English speaking operating in a completely French environment, which made misunderstandings a common occurrence even when ordering breakfast.

That visit was followed by one to the Rideau Waterway, where another hydrographic team was working in the vicinity of Smith Falls, within easy striking distance of Ottawa. Then to the Trent-Severn Waterway by MOT helicopter, where another survey was in hand around Buckhorn Lake near Peterborough. From there to the Upper Ottawa River near Pembroke, and finally a longer journey into northwestern Ontario to a major survey in the Lake of the Woods. I had now met many of my hydrographers-in-charge and their staff and crews and was impressed with their work ethic and ability to face adversity. HiCs such as Gerry Wade, Earl Brown, and Bruce Wright in particular made a mark. Gerry Wade was an older no-nonsense engineer who had seen battle in Belgium and the Netherlands and ruled his party like a hard-nosed sergeant major. Earl Brown was a product of the Northern Alberta Institute of Technology like Neil Anderson, and full of ideas to improve hydrography—a most efficient leader. Bruce Wright came from a land survey background, and sounded a bit aggressive when I first met him, but I soon realized that was his way when he first met people. His crew respected his capabilities and worked hard for him in a most challenging area, the Lake of the Woods.

In July 1969, I traveled to Tuktoyaktuk to inspect our Polar Continental Shelf Project (PCSP) operations in the Beaufort Sea. PCSP had established a permanent land base at Tuktoyaktuk, and operated a Decca 6f chain covering the Beaufort

Sea from Herschel Island in the west to beyond Liverpool Bay in the east. While in Tuk I paid a courtesy call on Charlie McIntosh, who was now in command of CSS *Richardson*. It was then on to Herschel Island by light aircraft to visit with our hydrographer-in-charge, George Yeaton, and the Decca Chain station on the island. George made me very welcome and contrived to persuade me to stay by displaying his best culinary skills. I was pleased to note that the chain would be available for use by all vessels transiting the area, including *Richardson*, which would now have a backup system to her Accurate Ranging Radar Unit and transponder beacons.

The one remaining major operation to be viewed was the facility at the Canada Centre for Inland Waters in Burlington, Ontario, with its fleet of ships and launches and workshops. I found myself being pulled down to CCIW on numerous occasions to sort out jurisdictional problems that had developed between my staff and the staff with operational responsibility for scientific studies on the Great Lakes and other adjacent inland waters. Providing vessel support to a large group of scientists, with varied interests and often little understanding of logistics, was a difficult and demanding task.

On top of all the foregoing, I now had to visit CHS Headquarters in Ottawa more frequently, where the new Dominion Hydrographer, Art Collin, was now at the helm and asserting his authority and power. It was an interesting situation, where I was responsible for the region's operating budget (hydrographic and fleet) and reported directly to the Director General Marine Sciences, but I reported to the Dominion Hydrographer on hydrographic policy matters such as nautical charting priorities and the development of better-trained and educated hydrographic staff. It was not an easy relationship, but the DH and the Regional Hydrographers gradually developed a good rapport. The first intermediate course was an unqualified success, and a university-level training plan soon followed, with half a dozen likely members of CHS being immediately approved by the Management Board for tuition and living expenses. An annual interview and reporting system for hydrographic surveyors was established, with HiCs being instructed to conduct serious reviews of their staff and to no longer opt for friendly comments on all. After many years of easygoing personnel reports, the change was not accepted with enthusiasm by all, but nevertheless it gradually became accepted practice.

A change of pace took place in the late fall of 1968 when the Dominion Hydrographer, Harvey Blandford, and I took part in a US conference held in New Orleans by the Institute of Navigation. There was a large element of hydrography in the conference, which drew our interest, and of course New Orleans was not a bad place to be in late November. It was also a good opportunity to get to know one another away from the constraints of the civil service. Mind you, ogling a naked female dancing around on the bar top above us could scarcely be supported as developing new hydrographic techniques.

Back in Ottawa I found myself attending ever more meetings at CHS headquarters, and I began to think seriously about obtaining official support for the idea of transferring all Central Region activities to the Canada Centre for Inland Waters in Burlington. A meeting with Bill Cameron, Director General Marine Sciences, confirmed his support for the idea, but he felt that the timing was not yet in our favour. Previous informal discussions with Al Prince, Director General Inland Waters, and Jim Bruce, the CCIW Director, nevertheless led me to believe that such a move might well be welcomed.

Meanwhile, I was getting to know my staff, and presumably my staff were getting to know me. The lack of Master Mariners in the Central Region hydrographic division continued to bother me, but during the winter we were able to recruit a couple of well-qualified people. Nationally there was not much support by CHS Management for such recruitment, with most hiring being from Land Surveying Technical Institutes or, where possible, recipients of specialist university degrees for tidal prediction or positioning and acoustics experts to enhance our in-house research and development activities. I was afraid that we lacked balance for the long haul.

My activities as president of CHA finally brought me into conflict with my duties as a member of the CHS management team. I felt that I had steered a fairly neutral path between these two levels of authority, but other members of the management team felt that when it came to selection of staff for university training assistance, I should not be wearing two hats. I reluctantly concurred.

The next Hydrographic Conference was held in Victoria in early March 1969. It was a first, and turned out to be a well-planned and well-run affair. The vessel CSS *Parizeau* was on display after her successful work in the Western Arctic, and the new Bertram launches, built locally at Canoe Cove, were put through their paces. These developments owed much to the input of Ernie Geldhart, the marine superintendent, and to Stan Huggett, tidal surveyor and Master Mariner, who perfected the working deck design of the *Parizeau*. The CHS management team took advantage of their presence to conduct interviews with Pacific Coast staff, while the CHA held somewhat heated meetings to elect a new slate of officers. Personally, it was good to be back in Victoria even for a short spell. There was no place I would rather be. The usual last night banquet and dance produced a very subdued gang of hydrographic surveyors on the flight home to Ottawa.

Adam Kerr, Master Mariner, who had commanded *Cartier* previously, was now the head of the R&D group in Central Region while he attended Carleton University seeking a degree in computer studies. This talented fellow, who could be very persuasive, now proposed to me that a regular hydrographic survey planned for the mouth of the Saguenay River in 1969 be turned into a large demonstration survey of all the R&D work that was presently being conducted in each region. It was a very interesting proposal, but could hardly be justified on a "dollars and cents" basis. After all, production was the name of our main game, and we needed those charts of key areas such as the confluence of the St. Lawrence and Saguenay Rivers. I therefore decreed that he could have his exotic demonstration if he married it to a regular ongoing survey.

Organizing and running such a massive undertaking took much time and the energies of a number of dedicated individuals. Central and Atlantic Regions contributed launches and equipment for testing, together with a Hovermarine craft specially brought over from the UK. Computers and their ancillary bits and pieces were everywhere, together with a small army of geeks and techies in support. In addition were, of course, the working crews and their launches and equipment dedicated to the regular ongoing survey. The base of operations was Tadoussac at the mouth of the Saguenay River, with most of the staff and crews billeted at the elegant bur elderly Tadoussac Hotel. Financially, it was a good deal for both hotel and CHS, with the hotel able to maintain near full capacity even in the off season, and CHS able to get very reasonable rates in comfortable and pleasant surroundings. Additionally, the food was very good, the waitresses were young and attractive, and the wine flowed of an evening.

The inclusion of the Hovermarine in the R&D portion of the activities was a contribution sparked by the Dominion Hydrographer himself, who found the money to bring such an exotic beast over to Canada. The hovercraft came with a small crew and support staff, a rather rambunctious gang who made sure that life was never dull. When I invited the DH to visit and involve himself in the activities, he was delighted to do so and spent some time on board the Hovermarine. At three o'clock the following morning he was awakened by a phone call from an unknown person with a strong English accent, who wanted to know that now that he had a hovercraft, was he perhaps interested in buying a battleship? He was not amused, but never suspected that the call came from one of the Hovermarine crew, who had just finished his daily report to the UK (five hours time difference) and probably felt that if he was awake, everyone else should be as well.

Although in the main things went well, there were the inevitable mechanical and electronics breakdowns, often taking place in the turbulent tidal waters at the junction of the St. Lawrence and Saguenay rivers, with the added hazard of fog ever prevalent. There were sometimes anxious moments when smaller vessels had to be rescued and towed back to safety. Weekends became periods of relaxation, and odd happenings occurred in the night. Once staff awoke to find that an office desk has been hoisted up the flagpole, a silent witness of some altercation between the CHS and the hotel staff.

Adam Kerr, as HiC of all he surveyed, appeared to be thoroughly enjoying himself, and not forgetting to pay close attention to some of the more attractive female staff in the hotel. Joe McCarthy, a junior hydrographer, was his quartermaster and doing a great job of keeping supplies flowing while fending off the more insistent demands of the many prima donnas who had flocked to take part in the exotic affair. A young distant relative of mine from England, Bernard Cullen, was serving as a crew member and thoroughly enjoying himself. I warned him about the local girls, but he just laughed and said that he had no attention of getting married and settling down. He eventually escaped back to Cambridge University unscathed!

The visits to the Tadoussac area were also an opportunity for me to maintain contact with other federal government agencies in Quebec City and Montreal. The St. Lawrence Seaway Authority extended its jurisdiction over a vast area, all requiring up-to-date Canadian Hydrographic Service nautical charts and supporting publications.

Our other field parties were active throughout Ontario, with particular emphasis on the Rideau, the Ottawa, and the Trent-Severn Waterways. Additionally, our survey of the Lake of the Woods proceeded apace. I visited them all, enjoying myself thoroughly in the fresh air and comradeship of my staff. These were happy days.

Interesting events were taking place in the Western Arctic, where the region had its strong connection to the Polar Continental Shelf Project. It became of national concern when the large US tanker *Manhattan*, escorted by the Canadian Icebreaker *John A MacDonald*, traversed the channels and seas of the Northwest Passage. After clearing the Prince of Wales Strait and heading westward across the

Beaufort Sea, *Manhattan* took the lead and thereby precipitated the famous incident of the "Admiral's Finger."

On *Manhattan*'s echo sounder, the bottom suddenly shoaled rapidly, causing the HiC Ken Williams to draw the attention of Admiral Storrs, Commissioner of the Canadian Coast Guard, to the alarming and unexpected development. Admiral Storrs poked his finger in the direction of the sounding graph and its depiction of an ever-rising peak, and somehow managed to damage the stylus and bring the echo sounder to a grinding halt—forever after known as the Admiral's Finger!

This was probably the first discovery of a submarine pingo in the Beaufort Sea, although some unusual shallow soundings had been noted on track soundings undertaken across the Beaufort Sea back in the fifties during the DEW Line construction and resupply missions. Many more were discovered after 1969. They are similar in shape to those dotting the coastal land mass around the Mackenzie delta. They have an ice core, probably "ice boil" describes it best, and are encased in dirt, rock, sand, and mud. Those with a minimum water depth of twenty metres or less are definitely a hazard to proposed supertanker passage in the Beaufort Sea.

Plans were immediately drawn up for a major surveying effort in this most important area in 1970, which would include our PCSP hydrographic survey group utilizing hovercraft in the inshore off Herschel Island and further east in the vicinity of Cape Bathurst and Darnley Bay. The PCSP shore base at Tuktoyaktuk would serve as a coordination centre for all seagoing and land-based projects being planned for the Beaufort Sea in 1970.

Meanwhile, several significant events had occurred in Ottawa. The Dominion Hydrographer had been accepted at the National Defence Staff College in Kingston and would be available for duties in Ottawa only on an opportunity basis. Colin Martin would be the Acting Dominion Hydrographer for the next year. Art Collin traveled to Malaysia while on his Staff College assignment and was entertained by my brother Gordon and shown the sights and delights of Kuala Lumpur. Knowing my brother, I could well imagine the grand tour upon which both of them embarked. The visit to Kuala Lumpur coincided with a large-scale interracial riot that shook the confidence of the government.

I finally managed to get Bill Cameron's approval to move Central Region out of Ottawa. He was most emphatic that Marine Sciences keep control of the fleet based at CCIW, indicating that the Inland Waters Directorate had been pushing hard to have the fleet under their control. My proposal to move the entire region out of Ottawa in the summer of 1970 was approved, and the decision conveyed officially to the Inland Waters Directorate. Now all I had to do was persuade my eighty or so Ottawa-based staff members that it was a good idea. It was a tough sell, as many members had deep roots in Ottawa, or were afraid of the unknown in moving to a new area. If it had been left to a free vote decision, I am sure that the region would still be based in Ottawa today. But this was not a democracy, it was a government operation, and it had to adjust to the dictates of logic. My salesmanship finally persuaded most key personnel, and most of the others reluctantly fell in line. For family or other urgent reasons, a small number elected to try to find other employment in the civil service in the Ottawa area. The die was cast, and the planning could begin. But I was learning fast that power and popularity do not go hand in hand.

Chapter Twenty-Five

A New Base – The CCIW Fleet – Searching for Pingoes

Early in 1970 I brightened up my office with a new personal secretary. Her name was Lakshmi Ram, and she was from Delhi, India, where she had worked at the Canadian High Commission. She was dark and exotic, with a good sense of humour. She had a university degree and was intelligent and articulate in English. I suspected that she was a bit lazy from time to time, but her other assets made up for that deficiency. I was glad to have her on my team moving out of Ottawa.

Figure 60 - Canada Centre for Inland Waters, Burlington Ontario, 1970

Unfortunately, the facilities manager at CCIW was unable to complete arrangements for our move into CCIW until later in the year, so alternative arrangements were made to move into an empty federal government premises in Hamilton for the interim period. It was spacious enough but needed much upgrading to meet our requirements, which had to be deferred because of our short six-month lease. However, some key personnel were already being fitted into CCIW, and regional control of all marine sciences branch operations was now assured.

The fleet base at CCIW was a model of its kind, with dock space to enable three or more fair-sized research vessels to dock in safety. A protective breakwater ran offshore, ostensibly to protect the vessels from ice pressure building up in Burlington Bay. It was not long enough, however, to protect the launch basin at the western end of the wharf, and considerable damage could be expected when heavy ice and weather combined in the early spring to funnel into the basin. All smaller craft were lifted onto suitable pads onshore for their safety. The remainder of the craft in the basin, of the forty-foot variety, were of steel or aluminum construction, and deemed to have a better chance of surviving a late winter gale. There were concrete ramps that led to the repair shops, with a mobile hydraulic lift that could easily transport the launches from the basin.

The larger craft left in the basin were a mixed lot, often constructed for specific tasks, such as geologic sampling, collecting air and water data ahead of weather fronts, sampling biota, or conducting a whole range of activities required by a particular scientific programme. These launches were usually under the direct operational control of Barrie MacDonald, Master Mariner and hydrographic surveyor, who was the Inland Waters Directorate scientific support coordinator.

The Jewel in the Crown at CCIW was, however, CSS *Limnos*, a one-hundred-fifty-foot-long steel-hulled craft built at the Port Weller Dry Dock in 1968 especially for scientific studies of the Great Lakes. *Limnos* had a beam of thirty-one feet and draft of eight and one-half feet, and her extensive working deck was only three feet above the waterline. Her gross tonnage was four hundred fifty-four tons, propelled at ten knots by two geared diesel Caterpillar engines controlled through two directional propellers. She was fitted with a main Galise hydraulic crane and several smaller auxiliary cranes. Her workboat was a Boston whaler, while she had well-equipped wet and dry laboratories. Her navigation equipment was an Arma Brown gyrocompass, a Bridgemaster Decca Racal radar, and an Elac echo sounder.

Limnos had crew of fourteen, being made up of six officers and eight crew members. She was in great demand by the scientific community, and our proudest possession. In reasonable weather she performed well, but in rough weather, with her shallow freeboard, she could be a bit of a beast to sail. With careful handling and thorough annual refits, she should still be in good shape for many years to come.

Figure 61 - CSS *Limnos*, Great Lakes 1974

When I took over the managerial responsibility for the region, we had no marine superintendent in place at Burlington. Ship branch in Ottawa helped out in the interim by lending us a shore-based engineer, John Higgins, who had formerly been Chief Engineer on the *Baffin*. He continued to base himself in Ottawa, driving down to CCIW for a few days each week. Reluctantly, I had to take steps to bring in a full-time marine superintendent to strengthen the region. I picked Austin Quirk, who was a Master Mariner hydrographer from within the region with lots of experience in mounting launch party operations, and who was acceptable to the hydrographic and scientific community.

Central Region's hydrographic launches were Bertrams in 1970, and proved adequate for most projects. However, with the computer age upon us, we were already considering the construction of slightly larger boats that would have cabin space, well protected from the weather. For several reasons, CHS seemed unable to take a national approach to the problem, resulting in three distinct designs being developed for each of the Atlantic, Central, and Pacific Regions. I was all for decisions being made locally to meet local needs, but surely a national coordinated approach, coupled with strong local input, would have provided a basic hull that would serve as a form for all, with separate regional requirements taken into consideration. Surely it would have been more economical. I lost the battle.

Shortly before our family prepared to take up residence in Burlington, my wife had a spectacular win at the local church hall bingo. It was not huge, only five hundred dollars, but still a large unexpected bonus in those days. I still recall the event, as I was stuck in bed with a back injury brought about by my curling activities—she charged up the stairs and flung the money at me as I gaped in astonishment. We traveled down to Burlington without too much incident, almost losing the cat in Oshawa and then, upon arrival at our new home, finding that our very ripe garbage had been packed away with our furniture and personal clothing in Ottawa. It was then decided that the five hundred dollars would be used to provide the down payment on air tickets to enable Doreen and the children to fly out to Victoria for a well-earned holiday and rest with members of her family still living in Victoria. I could get on with managing a region in transit and also in transition.

I had also arranged to meet Mike Bolton, Regional Hydrographer Pacific, in Edmonton en route to Tuktoyaktuk and the large hydrographic operation about to take place in the Beaufort Sea. We flew PWA to Inuvik and then on to the Polar Continental Shelf Project base at Tuktoyaktuk. From there we traveled by helicopter to Darnley Bay, where PCSP were conducting hovercraft sounding trials across open water leads that had appeared between large floes of floating ice. The trials went well, but obvious weaknesses in both design and deployment of the hovercraft were soon apparent. In any sort of wind the hovercraft steered like a crab, and the noise level was absolutely deafening, making communication without ear protection almost impossible. We admired the hydrographer and crew for their perseverance, but decided that we had seen enough. Upon returning to the PCSP base camp, we were bundled aboard a small fixed-wing aircraft and flown westward along the Yukon coast to Herschel Island. When we had traveled by helicopter to Darnley Bay along the Beaufort Sea coastline, we saw very little wild life, but the Yukon coast in July was just teeming with life of all kinds. Geese, swans, ducks, and myriad other smaller birds were everywhere to be seen, and vast herds of caribou and reindeer moved across the verdant landscape. Flowers and shrubs of every possible colour and description were in full bloom, and only the inevitable mosquitoes and black flies were left to our imagination. We were sorry to leave the coast and approach Herschel Island, where we landed on a short sandy beach close to the old but still-standing RCMP post built back in the 1890s.

It was no longer an RCMP base but now contained a PCSP outpost fitted with a Decca 6f electronic positioning device and a powerful radio transmitter and receiver. The outpost was under the command of George Yeaton, a Central Region hydrographer. He made us very welcome while we awaited the anticipated arrival of CSS *Parizeau*, a Marine Sciences branch vessel based in Victoria, and a significant part of Mike Bolton's responsibilities in the Western Arctic. George was a good cook and good company, so we spent a pleasant day or so exploring the island accompanied by some of the odd scientists that based themselves at Herschel Island, supported by PCSP. They were interested in all kinds of fossils, birds, seals, and arctic char, but not in pingoes, which they treated with bemusement when advised of our concerns. It led to some interesting discussions.

Figure 62 - CSS *Parizeau*, Pacific Coast 1970

Eventually, *Parizeau* appeared out of the fog, and we were taken on board to meet the captain and the hydrographer-in-charge. The captain was Alan Chamberlain, an Englishman who had been with the South American Saint Line. The HiC was Stan Huggett, former Master of *Richardson* in 1966, and who had received his seagoing training and qualifications while serving with the Park Steamship Line. Faced with a working visit from the Brass, they consigned us to the depths of the vessel, where we lacked portholes and even prestige! However, the ship was full, so I suppose we had to be content with such substandard accommodation. They made up for their inability to honour us properly, however, by ensuring that we were well wined and dined.

Parizeau was named after the famed irascible hydrographer who dominated hydrographic surveying on the Pacific coast for so many years and maintained a largely independent role in defiance of the strictures promulgated by Ottawa. She was built in 1967 at the Burrard Dry Dock in Vancouver. She was constructed as a working hydrographic/oceanographic vessel with accommodation aft of a clear working deck and raised forecastle. She carried six officers and fourteen crew members. Her length was almost two hundred feet, with a breadth of thirty-nine feet, a draft of fifteen feet, and a freeboard of five feet. Her gross tonnage was one thousand, three hundred fourteen tons. Two Deutch geared diesel engines powered twin controllable-pitch propellers at a cruising speed of ten knots, with a maximum of fourteen knots if required.

Additionally, she was equipped with a bow thruster, Sperry gyrocompass, two Decca Racal radars, an Elac echo sounder, a large Jacobs crane and auxiliaries, lifeboat/workboat, and laboratories on both upper and lower hydrographic decks.

During our visit, we observed *Parizeau* undertaking sounding operations and tidal current studies and made arrangements to transfer to the CCS *Baffin*, whose arrival in the Beaufort Sea from the east was expected momentarily. She was the largest ship in the fleet, built in 1956 and equipped with four large sounding launches. (I first mentioned *Baffin* in Chapter 13.) She exceeded three hundred feet in length and had a cruising speed of twelve knots, with a maximum speed of fifteen knots. Her crew of ninety-two souls was augmented by a team of ten or twelve hydrographers during a typical season. For the "Pingo" operation, she was carrying four special high-speed launches jammed with the latest in hydrographic positioning and sounding technology.

Due to weather and time restraints, we were unable to make the transfer to *Baffin* and therefore had to settle for being returned to shore at the PCSP base at Tuktoyaktuk. The remaining ship involved in our Beaufort Sea project was the CSS *Hudson*. She was delayed rounding Point Barrow due to heavy ice around Point Barrow and did not appear in the Beaufort Sea until after we had returned to civilization.

CSS *Hudson* was primarily an oceanographic vessel, but especially refitted for geophysical work on the Pacific coast after traversing the Panama Canal. She had been built in 1963 at the Saint John New Brunswick Shipyard and normally carried a crew of thirty-seven souls. *Hudson* was two hundred seventy-five feet long, with a forty-eight foot beam, and a draft of twenty-one feet with a freeboard of ten feet. She displaced three thousand, seven hundred forty gross tons, being propelled by four diesel Alco engines at a cruising speed of ten and a half knots and a maximum speed of seventeen knots through twin-pitch fixed propellers. As with *Baffin*, she was fitted with hangar and flight deck on the stern. Her other equipment and fittings consisted of a bow thruster, a Sperry gyrocompass, two Decca Racal radars, an Elac echo sounder, a large Arva telescopic crane, and smaller auxiliary cranes. Finally, she had two laboratories, one hydrographic, one oceanographic.

Hudson was the key to the success of the project, which indeed was christened by the director general of Marine Sciences as "Hudson 70." He had become very much involved in the affair, traveling on board *Hudson* on her circumnavigation of North America, as far as Victoria, BC. He had traveled into the Beaufort Sea many years previously as a young scientist on a small, converted fishing vessel and was therefore most nostalgic and enthusiastic. The geophysical operations refit of *Hudson* was completed, and only the ice prevented her from joining *Baffin* in the search for the elusive pingoes off the Mackenzie delta.

The project was finally completed fairly successfully, with *Baffin* doing the lion's share of the work, and with many potentially dangerous pingoes discovered and delineated. However, *Hudson* was withdrawn to the east before her task was completed—an unnecessary panicky move dictated by an incorrect reading of changing ice conditions. Nevertheless, much had been accomplished by all three vessels and the PCSP hovercraft. Mike Bolton and I were able to hitch a ride on a small aircraft leaving Tuk for Inuvik. Its only other passenger was Commodore Robertson RCN, of the Icebreaker HMCS *Labrador*, who made us very welcome. He and Mike had been shipmates on the ship back in 1957 when they rendezvoused with the US Coast Guard Cutters *Storis*, *Bramble*, and *Spar* in Bellot Strait as they were completing their epic circumnavigation of the Arctic Ocean. A nice ending to a somewhat frustrating venture.

I returned home to Burlington shortly before Doreen and the children completed their long holiday in Victoria. They had all obviously benefited from the change of pace and looked tanned and fit. I was overjoyed to see them all—living a bachelor existence for a lengthy period was no longer the least bit attractive to me. I oversaw the move from our temporary quarters in Hamilton to our permanent base at the Canada Centre for Inland Waters. It was not entirely to our satisfaction, as Inland Waters Directorate were the landlord and we had to make do with accommodation scattered throughout the buildings. Hydrography was located on the ground floor beyond the hydraulics laboratory, with their all important compilation group fitted into space on the other side of the hydraulics lab but on the second floor. I was allocated an excellent office close to Jim Bruce, the IWD head, but a long way away from my ships division and R&D group to the east and my hydrographic division to the west. Even my administrative, financial, and personnel staff were located some distance away. We were a region now under one roof, thank goodness, but still lacking some cohesion.

Sandy Sandilands of Pacific Region joined us as Acting Assistant Regional Hydrographer during this period, and was a great help in making sure that the hydrographic division fitted smoothly into its new quarters at the Canada Centre for Inland Waters. It was also good for me to have someone around from the Pacific coast, and part of our shared troubles and triumphs. Harvey Blandford was on temporary assignment at headquarters in Ottawa.

I found my proximity to the IWD director at CCIW, and my relationship to him, a bit of a strain at times. He was a nice chap but determined to maintain a certain image: he was the director of CCIW and all it contained. This image was promulgated at length to the public through adroit public relations with the press and the media in general. I even found myself acting on his behalf when he was off on some other activity, and vainly trying to explain our complicated bureaucratic relationship to some irate politician who demanded a CCIW response to his complaint about some perceived impending environmental disaster. It required a tricky balancing act—ensuring independence of action while maintaining reporting and policy links to my own branch in Ottawa. I determined to ease the situation as soon as possible.

Figure 63 - CSS *Port Dauphine*, Welland Canal 1970

With family settled in place and children back to school, with football and dancing high on their agenda, my wife and I were able to resume a social life when my travels did not intervene. As mentioned earlier, I was a Rotarian, and in Ottawa had joined the Ottawa West club. I now transferred to the Burlington club, which met at the Burlington Golf and Country Club, not too far from my home. I also joined the Toronto Marine Club and transferred my Canadian Legion membership to the Burlington branch. Other involvements out of Burlington were with the Canadian Institute of Surveying, where I chaired the Hydrographic Committee. All in all, I was a busy fellow.

1970 ended well, with *Limnos*, *Port Dauphine*, and our chartered ship *Martin KaarLson* secured safely alongside the wharf at CCIW, suitably bedecked to celebrate the festive season.

Chapter Twenty-Six

Expanding Horizons – R&D and Europe

I have not mentioned previously the growing close relationship that CHS Central Region had developed with the US Great Lakes Survey, formerly under the control of the US Army Corps of Engineers, but now directed by the US Coast and Geodetic Survey. Their first director was a Captain Bob Williams with whom I established good rapport, which led to excellent cooperation on matters of international concern, and eventually to an exchange of technical personnel to learn from one another the best of our methods. This approach was soon adopted nationally by both nations, and a more formal arrangement was set in place. Bob Williams retired to the Seattle area and was succeeded by Captain Ken MacDonald, who took a big part in the planning and operations on the US side of the International Field Year Great Lakes (IFYGL), which was the big event of the 1971 season.

In addition to lengthy planning meetings at CCIW, I traveled several times to Washington, D.C., to finalize the planning for IFYGL, where my main task seemed to be to persuade my US counterparts that the many vessels taking part in the event needed a common system for positioning on Lake Ontario, where most of the activities would take place. We would have liked to experiment with the most modern positioning systems available, but time and limited funding led us to purchase a version of the already tried-and-true Decca 6f system from the UK. Arrangements were made for all vessels and aircraft taking part in IFYGL to be equipped with the positioning system and crews instructed in its operation. I might add that I also had to persuade CCIW scientists of its necessity to the success of the project. I was becoming ever more aware of a stubborn mind-set that made negotiations with many in the scientific community at CCIW a difficult undertaking.

At the height of IFYGL, the wharves at CCIW were kept very busy with Canadian and US vessels refueling, changes in staff and crew, and huge amounts of data turned over to be processed by the CCIW main computer facility. In the middle of all this marine activity, the region acquired a new vessel, the former personal vessel belonging to the estate of John David Eaton of the well-known Eaton Department Stores family. Prior to formally acquiring her, I traveled to Toronto Harbour, where she was docked, to assess her worth. She was a beautifully maintained ship of about one hundred and eight feet in length, with adequate space for a working crew and laboratory space. She had been employed as an oceanographic vessel by Norwegian interests before becoming the personal yacht of John David Eaton, who deployed her mainly in the tropical waters off southern Florida. Her owner had died, and the Eaton family wanted to dispose of the ship as quickly as possible. I said that I would like the ship if we could find the money in our budget. I have never seen, before or after, such quick action by government to enhance a sale. She was acquired for the sum of $250,000, a bargain even for those days. The minister of the Treasury Board, Bud Drury, was a personal friend of the Eaton family and expedited the sale so that she became part of our fleet within three weeks of my saying the word "Yes"! A remarkable affair and an illustration of the links that bind the powerful—political and monied.

Figure 64 - CSS *Bayfield*, Burlington Bay 1970

When she became the latest addition to our fleet, she was immediately christened *Bayfield*, after Admiral Bayfield of the Royal Navy, who did much surveying in the Great Lakes before going on to complete the surveys of the St. Lawrence River, Prince Edward Island, and parts of Newfoundland. She was not our first *Bayfield*; the original had been the former tug *Edsall*, acquired in 1884, which was refitted for hydrographic operations on Georgian Bay and given the honoured name. A second *Bayfield* followed with the acquisition of the *Lord Stanley* in 1910. This third CSS *Bayfield* came to us complete with a full set of spares for all equipment on board, together with silverware and linen. The only item retained by the Eaton family was the contents of the large and expensive wine cellar. What a shame!

Then, unexpectedly, I found myself preparing to go to Europe to attend the Commonwealth Survey Officers Conference in Cambridge and to present a paper on behalf of the Dominion Hydrographer at the International Federation of Surveyors (FIG) Congress being held in Wiesbaden, West Germany. The trip was tied in with a visit to Decca in England related to our purchase of the Decca 6f system for IFYGL and visits arranged to a number of establishments in the UK where advanced research and development work was underway in side scan sonar and associated systems.

I met with the Dominion Hydrographer in Cambridge at the Commonwealth Survey Officers Conference, where we discussed his paper on the Canadian Hydrographic Service that I was to deliver in Wiesbaden. I concurred with most of the points he highlighted in his paper, but quietly decided to make my own views known if questioned by the chairman or members of the audience.

The Cambridge conference was interesting in that it brought together surveyors from around the Commonwealth in their different disciplines, but it also featured a number of Americans, who took part in the proceedings as if their country were still part of the old British Empire. The setting was impressive in the old Guild Hall, with many of us quartered in Pembroke College, where we downed treacle pudding for lunch, washed down with many a tankard of ale. It was a wonderful experience, and ambling through the college gardens afterward one could feel the presence of the past history of this land. Meeting other surveyors was a good experience, but the calibre of papers delivered was definitely a mixed bag. I felt that the paper selection board had been democratic rather than demanding.

Off then to London to conduct some family business. With the help of my niece Micky, who was working in London, I managed to locate Mrs. Luckman, an elderly lady who had been my great-uncle Duncan's housekeeper back in my seafaring days (see *Mandalay to Norseman*). She was now quite frail and a bit confused, but we managed to have a conversation about the past. It was then time for a little relaxation. Neil Anderson of CHS was to be with me for the remainder of my European jaunt, so I decided to impress him by setting up a date for us with two beautiful girls. Micky contacted Maria, another youthful relative of mine, and both these young ladies, well endowed and dressed in their finery, reported for duty at our hotel. Neil's eyes were a sight to behold—he was obviously impressed by the ladies and also by the hidden talents of his companion. Of course, I had to confess to skulduggery in supplying these delectable females on such short notice. An excellent dinner followed by some mild dancing ended a very enjoyable evening.

Neil then went off to visit some government establishment outside of London while I traveled overnight by train to Glasgow to see my Aunt Bessie. She made me very welcome and arranged that I could attend a football match at Ibrox Stadium, the home of the Glasgow Rangers.

The game was a bit of a disappointment, but being there in person was well worth my while. Although I was in a more select part of the stadium, even so, the drinking that accompanied the match was truly astonishing. Meanwhile, other parts of the grounds were tightly patrolled by the police, and bottles and other objects were hurled at the opposing fans. I had forgotten how outrageous a Scottish football match could be. I followed the crowd departing the stadium and ended up in a noisy, smelly pub, where strangers did not seem to be welcome. I hastily found my way back to my aunt's and peace and quiet. I was obviously out of touch with the Glasgow scene!

It was very pleasant staying with my aunt and catching up on family affairs. With some regret, I headed south on the London morning train to immerse myself in taking a good hard look at some of the newest hydrographic technology being developed in the United Kingdom. I joined Anderson in Taunton, the headquarters of the Royal Navy hydrographic department. We made ourselves known to the Commanding Officer and were given a short tour of the nautical charts facility before meeting the man whom we had really come to see. He was Roger Cloet, a Belgian scientist who was doing some remarkable work with side scan sonar on bottom sand waves in the Thames estuary. His studies clearly showed the migration of sand banks in the vicinity of navigation channels, a matter of some concern. We then visited the Decca plant at Leatherhead to see their latest positioning devices, while I concluded our arrangements regarding the deployment of their positioning chain in Lake Erie. From there it was on to Great Yarmouth and Lowestoft in East Anglia to witness the display of a sonar mapping device that had detected many previously unknown vessel wrecks off the coast. It had been developed by the National Fisheries Research Establishment and did seem to show great promise for the provision of bottom mapping coverage.

It was now time to catch up on our note keeping and prepare for our journey to the Continent. We caught a modern fast ferry from Harwich that took us fairly speedily across to the Hook of Holland. I managed to get myself a cabin and relaxed in comfort as we made our way through a maze of sea-going traffic wending their way through light fog that kept their foghorns blasting almost continuously. At the Hook of Holland, there seemed to be a veritable armada of dredgers heading seaward to discharge their cargoes of tailings and returning to continue the cycle.

We sleepily made our way to a train that deposited us in The Hague on a rainy evening. Our hotel was small but nice and neat, and the Dutch people were all very friendly. The following morning we visited the headquarters of the Rykwaterstaat—the equivalent to our Public Works department. They were responsible for all works on the inland waters of the Netherlands, including hydrographic surveying, and were carrying out some interesting work involving underwater positioning and mapping. We were given a grand tour and then fed heaps of Indonesian rijsttafel, washed down with copious amounts of Amstel beer.

As we were leaving for Germany the following morning, we decided to have a tour of the sights of Den Haag. These included a small red-light district with ladies of the night on display and shops with displays of sexual appliances and toys. All this in full view, only a short distance from the bustling downtown city core. The Dutch were truly a remarkable people!

Our train departed on time, heading toward the border with Germany and passing through Nijmegen of wartime fame. We were then boarded by German police and customs before continuing on our way into Germany. Having seen many a film of wartime days with desperate individuals trying to outwit the Gestapo, I looked at these intruders with some suspicion. However, they gave us no trouble, and very shortly thereafter we were heading down the west side of the Rhine River toward Cologne. That city had, of course, been very heavily bombed in World War II, and signs of the devastation were still apparent. In particular, the mighty Cathedral, while dominating the city, was still under extensive repair. We then continued our journey southward alongside the river, coming into the land of castles, fortresses, and extensive vineyards climbing the slopes high above. A truly wonderful sight.

We departed the train in Frankfurt, and traveled by bus to Wiesbaden, the site of the 1971 FIG Congress. Wiesbaden was a spa city, with many fine buildings and a magnificent casino. It was a far cry from what I thought Germany might look like. When I left the UK to emigrate to Canada in 1948, Germany was still prostrate and largely devastated in the aftermath of her defeat. Now in 1971, the signs of recovery and prosperity were evident all around me. It took a bit of getting used to.

Our hotel was an old-fashioned building, filled with heavy, dark wooden furniture, carvings, and statues. However, the claustrophobic atmosphere was eased somewhat by a small balcony that served both our bedrooms. This also came in handy when the Rhineland experienced an unprecedented heat wave, and quaffing of white wine and good beer became almost a full-time occupation The food was a bit heavy for the climate, but the pork and assorted sausages were not to be sneezed at. Right next door to our hotel was a sidewalk café much occupied by a small troop of the best-looking whores I have ever seen. They looked like they had stepped out of the chorus line of a West End London musical extravaganza. They even caught the eye of a dapper older man passing by, who doffed his hat in appreciation. I suddenly recognized him—he was Rear Admiral Steve Ritchie RN, the former Hydrographer of the Royal Navy, and the chairman of our Hydrographic Commission.

Steve Ritchie was, at this point in his career, lecturing at Southampton University and starting another career as an author. He went on to become president of the Directing Committee of the International Hydrographic Bureau in Monaco. I made myself known to him, and we walked to the conference hall to register and get acquainted with the many other delegates there from around the world. The ladies of the sidewalk café gave us a resounding whistle as we marched off.

Figure 65 - Rear Admiral G.S. Ritchie RN CB DSC, Chairman, Hydrographic Commission, FIG, Wiesbaden 1971

As I mentioned earlier, this was a congress of surveyors from around the globe and covered every aspect of the surveying profession, from cadastre and land management and registration, to survey instruments and methods, to engineering surveys and surveying education, to professionalism and ethics, among many other matters of concern. Only very recently, at the 1968 Congress in London, was hydrographic surveying dealt with as a separate discipline, as its importance dawned on the private sector. Steve Ritchie had headed up the newly formed hydrographic commission of FIG in London, and was here in Wiesbaden to make sure that Commission Four, as it was now designated, made its mark.

FIG—the Fédération Internationale des Géomètres—was formed in 1878 in Paris by the national professional surveying associations of the seven leading European nations of the day. In 1971 in Germany, there were in excess of thirty national member associations from around the globe grouped in nine commissions dealing with various aspects of surveying, while engaging in the exchange of technical information that would benefit the profession as a whole, whether governmental, private practice, or academia.

There were seemingly endless speeches as the Congress got underway, with everyone congratulating the Germans on their efforts to make the affair a success. Germany had been a founding member of FIG in 1879, but the war years had affected its relationship with other members, and only now was Germany fully accepted back into the fold. It was also her chance to show off her material progress since the defeat and chaos of 1945. The Congress was of course conducted in the three official languages of FIG: English, French, and German, with simultaneous translation facilities provided.

It was with some relief that we adjourned to our technical deliberations. Commission Four had prepared a programme of eighteen papers to be given in six sessions, with time set aside for questions and discussions. There were papers dealing with the measurement of the velocity of sound in seawater, the metrication of charts, special charting needs for deep draft vessels, automated chart production, surveying on the ice in the High Arctic, and many others. However, the most significant paper, other than the one I was about to deliver, was a paper by Alan Ingham of the North East London Polytechnic on "Surveying Education for the Sea Surveyor." Alan was a former hydrographic surveyor with the Royal Navy, who now lectured on hydrographic surveying at NELP to a broad range of civilian surveyors now operating in the sea environment. His views were refreshing and most obviously applicable to much of what we were attempting to do in Canada, particularly in the Canadian Hydrographic Service. He was the author of a fine book on the same subject. I found out later that he was an exceptional watercolour artist, specializing in landscapes of his home county of Yorkshire.

Neil Anderson delivered an interesting paper on the techniques developed by PCSP to sound through the Arctic ice, and then it was my turn. I had hoped to illustrate the paper about modern techniques in use in the Canadian Hydrographic Service with brand-new film depicting our fast launches and other exotic pieces of equipment in action, but the film that arrived was not in a format that could be displayed on the available projector. It was therefore up to me to deliver the Dominion Hydrographer's paper in the fifteen minutes allowed. The paper was well received, and I had a chance to expound some of my own ideas and preferences. Steve Ritchie, as chairman and as a product of the naval mariner/hydrographer combination, and knowing of my own Master/

hydrographer experience, wanted my views on the general approach of the CHS where the roles of Master and hydrographer-in-charge were separated. I did not hedge my answer and indicated that if I had my way in the CHS, the roles would be combined. I hastily added that these were my own views and not those of the Dominion Hydrographer. A short discussion on the infamous Black Rock case then took place, where it could be said that the dichotomy of command led to the grounding of CSS *Baffin* on a hazard visible to all. My personal ordeal was over, and I could now enjoy the rest of the Congress.

There were many activities of a social nature, as many delegates had brought their wives and family along for the occasion. One real highlight for me was a cruise on the River Rhine, where we traveled by bus to Maintz and boarded a large ferry about nine o'clock on a Sunday morning. The craft was well appointed, with lots of seating and access to restaurants and bars. Our voyage took us northward through the castle country that we had glimpsed from the train en route to Wiesbaden. It was truly fairy-tale country, with the castles and keeps dominating the hills and the river channels. These castles and keeps must have collected countless wealth over the centuries, because their owners controlled the river traffic and extracted their tithe. Every so often we would dock at the pier of a small port, where could disembark and climb up to a partially restored castle, or just collapse at the nearest tavern to be soothed by quaffing the local brew. At one such pier we were met by an entourage in support of a Rhine Maiden who bore a huge goblet of sparkling white wine that she presented to the president of FIG, while allowing herself to be bussed by all those surveyors lucky enough to be close by.

We carried on down the river to the vicinity of Bonn, the capital of West Germany, and Koblenz at the mouth of the River Mosel. I found this area particularly interesting, as my father had been stationed at a place called Bad Godesburg when he served in the British Army of Occupation in the Rhineland after the Armistice of 1918. Indeed, there were paintings of the area on the walls of my family home in Greenock, Scotland. The place was also a favourite spa of Adolph Hitler in the mid thirties.

At one of the Hydrographic Commission sessions, I met Hans Ermel of the German Hydrographic Department located in Hamburg, who very kindly invited me to visit their establishment after the Congress ended. With Steve Ritchie's encouragement I agreed, while also making an appointment to meet with another Congress participant, named Wentzell, who represented the Atlas Echo Sounder Company. The final business of the commission was to thank Steve Ritchie for his contribution in bringing the hydrographic surveying profession into FIG, to recommend to the FIG Permanent Committee that Marc Eyres of France be the next chairman of Commission Four, that Dr Fagerholm of Sweden take on the duties of vice chairman, and that Gerry White of the Port of London Authority continue as Commission secretary.

The flight from Frankfurt to Hamburg by Lufthansa was uneventful, although I remember being astonished at the rush of passengers to pick up their plastic-wrapped snacks before boarding the aircraft—a new experience for me. Herr Ermel had booked me into a downtown hotel close to the main railway station. It was a truly appalling choice, as I soon found out. The rooms were below standard, and the place was crawling with bedraggled females wandering the halls looking for custom. Even a seaman of my experience was appalled. I was too tired to go looking for another establishment, so I locked the door to my room and headed out into the night. I decided to go to see the Reeperbahn in the St. Paulie district of Hamburg, close to the docks. I had never been there, but had had many descriptions of the place from shipmates.

First of all, I could not find any transport—the train schedules posted in the main station bewildered me and the buses were beyond me—and I was finding it difficult to make myself understood in the babble of German issuing forth from each citizen that I accosted. So, after getting an idea of the general direction of the docks, I set off from the well-lit railway station into the pitch-black night. I was moving through an area that had been flattened and burned by our bombers in WW II and had not yet been rebuilt. I recall passing the gaunt remains of a huge church or cathedral and feeling some pangs of remorse, but I soldiered on until I could see the bright lights of the Reeperbahn in the distance.

There was light, music, gaiety, food, and liquor in copious quantities side by side with squalor and degradation, all crammed into about ten blocks of older buildings. Everything and everybody appeared to be for sale, from the painted whores behind the candle-lit windows of the houses to the younger dancers plying their trade in the so-called night clubs. I wandered around soaking up the incredible atmosphere, sampling the good German sausage washed down with better German beer, and avoided or ignored the blatant actions of the painted ladies. One thing in particular intrigued me: credit cards were accepted for every activity, including those provided by the girls. I could not help but wonder how husbands explained away an American Express bill for the services of "Marlene" or "Lucy" to vigilant spouses.

I found a taxi eventually and was wafted back through the deserted streets to the centre of Hamburg and my dreadful hotel. I survived the night and went to meet with Mr Ermel and the officials of the German Hydrographic Service. I was given the grand tour of what appeared to me to be a very efficient organization with lots of scientific and technical talent, but with little appreciation as to how much the world of hydrography would change in the immediate years to come. However, they had arranged for me to board a Port of Hamburg catamaran that was fitted with a full sonar sweep boom extending fifty feet on either side of the vessel. The sonar equipment was manufactured by the Atlas Company, and was performing to specification standards. I was quite impressed. The vessel was the *Deepenschriever 11*, specially constructed for operations in Hamburg Harbour. The captain of the ship exuded an air of thorough competence.

I finished my fleeting visit to Hamburg being entertained to dinner by Mr. Wentzell of the Atlas Company. It was a fine dinner, with a wonderful Dover sole washed down by a very good Mosel white wine. Mr. Wentzell pressed me hard to buy his product, but I pointed out that we would require further demonstrations of the Atlas echo sounder in Canada before committing ourselves to a purchase. I headed back to Canada, grateful for the fine meal but leery of pushy salesmen.

Chapter Twenty-Seven

Changes in the Management Structure – Becoming a Bureaucrat

I enjoyed my sojourn in Europe but was glad to return home to Burlington in early September of 1971. My wife had soldiered on without me, coping with our noisy, irrepressible clan with considerable skill. I had to reconnect with them all, even if it was just to remind them that I was their dad. Job-wise, things were going along fairly well, although I was feeling the strain of being both Acting Regional Director and Regional Hydrographer more and more. It was particularly vexing because I was not being paid anything extra to occupy both positions. I resolved to have the situation clarified at the earliest opportunity.

IFYGL continued to do well, with the Central Region Ship division under Austin Quirk doing an excellent job of meeting the diverse needs of the multi-flagged fleet. A successful CHA/CHS hydrographic surveyors Intermediate-level course was held in the region. The hydrographic field operations being conducted on the St. Lawrence River at Matane, at Two Mountains, and in the vicinity of Cornwall were all going well. Additionally, we continued to chart the Rideau waterway and the Trent-Severn Canal, together with further work out at the Lake of the Woods. Our Tidal section was busy on the St. Lawrence River, and our R&D group were experimenting with side scan sonar applications both in the High Arctic with PCSP and in more southerly waters. All in all, we were a very busy region.

The news from the Beaufort Sea was encouraging, as the pingo field became almost completely delineated. Major operations continued out of the Bedford Institute in Nova Scotia, with some interesting work being carried out by Mike Eaton and his group on positioning by satellite and on radio controlling unmanned craft of various types.

The oil industry in Canada sponsored a hydrographic conference in Banff in October, which was attended by many of us in the government service. The Canadian Hydrographic Association was active in Alberta with the formation of the "Prairie Schooner Branch," a play on words depicting the old covered wagon from the west setting off on a sea voyage. The conference took place in the Banff Centre for Fine Arts, located amid the mountains outside the city. The oil industry was enthusiastic about hydrography and was becoming heavily engaged in the Beaufort Sea and, of course, off the Grand Banks of Newfoundland on the Atlantic coast. I was particularly glad to see development, as I had been encouraging land survey companies to expand into hydrographic surveying.

Our contingent were staying at the old Rim Rock Hotel, high on the side of a mountain. It was a sight to behold—the early morning sun climbing up over the rim of a spectacular rock face each day. All sorts of wildlife roamed close to the hotel, including elk and deer, and signs of bear could be found each morning on the trunks of cars containing food, where irate animals tried to force the trunk lids open. The Rim Rock today is a great modern edifice, where I am sure bear, elk, and deer are kept well away from the pampered guests.

Meanwhile back in Burlington, a number of changes had taken place, with Nelson Freeman taking the place of Brian Tait as head of Tides & Tidal Currents, and the region had taken additional responsibilities for Coastal Zone Mapping of the Great Lakes. This had been a joint provincial-federal responsibility that had almost been abandoned due to lack of funding support. The Inland Waters Directorate pleaded for its continuance at international meetings on Great Lakes Water Quality, and I agreed to see if the mapping could be integrated with our regular coastal surveying responsibility. Bill Haras, a coastal zone engineer, was transferred to the region to head up the programme.

The acquisition of Nelson Freeman as Chief Tidal Officer was to be of some significance to the region, as will later be shown. He had studied Oceanography at the Massachusetts Institute of Technology and was interested in expanding his tidal expertise into the broader field of physical oceanography, which coincided with my long-term aspirations for the region.

There were indeed major changes in the air. The new department of Environment took on the responsibility for the water sector, including the fleet. We were now part of a new sector known as Fisheries and Marine. The world was changing, but not all that much—fifty years previously we had been part of Marine & Fisheries. Where would we end up next?

The changes were more evident in Ottawa, where we were no longer an integral official part of the larger surveying and mapping community, but relied on ministerial letters of agreement to continue our cooperation in such areas as use of the printing facilities for chart production and in our R&D ventures with the Polar Continental Shelf Project. Energy Mines & Resources fought hard to retain its budget and overall position, and Environment Canada likewise pushed its needs with great vigour. We did not realize it at the time, but we were seeing the ending of the period of expansion that permeated the Government of Canada during the sixties and into the early seventies.

1972 dawned with the promise of further change in the air. I had accepted the professional responsibility of representing the Canadian Institute of Surveying as their delegate to the Hydrographic Commission of the International Federation of Surveyors. This duty entailed coordinating the Canadian response to our involvement in the 1974 Washington Congress and ensuring that the Hydrographic Committee of CIS supported our activities. Canadian papers were planned to cover aerial inshore hydrography, Rho-Rho Loran C for precise long range positioning, automation developments for marine chart production, and my own submission on Great Lakes water quality and the requirement for hydrographic surveying. I was being kept busy!

Bill Cameron had retired as Director General Marine Sciences, and Art Collin had been appointed in his place. Gerry Ewing, the Assistant Regional Hydrographer Atlantic, became the Dominion Hydrographer. I became director of the region, leaving the position of Regional Hydrographer Central Region vacant. It was hard to keep track of the changes underway, and who was doing what. Eventually, there was a competition for the position, which was won by Adam Kerr of Central Region.

Our acquisition of CSS *Bayfield* was proving to be worthwhile, as the Tidal division found her to be very compatible for the tidal current studies that the region was undertaking in the St. Lawrence River. An expert team was being created, and much useful data was collected, while new data acquisition techniques were developed. She was also deployed in support of the Olympic Games yacht races at Kingston, Ontario, and on a number of projects on the upper Lakes. Additionally, *Bayfield*, being formerly a private yacht, was well appointed to conduct official hospitality on board when the occasion demanded it. She was a fine little ship, looked upon with pride and affection by us all.

We acquired another vessel in 1972—she was the seventy-five-foot-long fast cutter *Advent*, built by Alloy Manufacturing of Lachine, Quebec, to our specifications. She was constructed of aluminum and displaced seventy-two tons gross. Her crew complement was made up of Captain, Engineer, and two deckhands. *Advent* was fast, being powered by twin diesel Detroit engines attached to twin propellers, with a cruising speed of sixteen knots and a maximum speed of twenty-two knots. It had sleeping quarters for four but was mainly used as a day work boat. The design was similar to that used by the oil rig supply companies in the Gulf of Mexico. An A frame was mounted on the stern for diving operations and sampling work. She was also fitted with a ten-ton crane on her ample rear deck. Her navigational equipment consisted of an Anschutz gyrocompass, Decca radar, and Raytheon echo sounder. Additionally, she carried a Zodiac work boat and contained a fully fitted hydrographic laboratory immediately abaft the wheelhouse accommodation. She became a well-loved ship.

Figure 66 - CSS *Advent*, fast cutter, Burlington 1972

In addition to the usual inspection visits to hydrographic field parties, two events of that year 1972 stick in my mind. It was the year of the ice hockey clash between Canada and the Soviet Union. The senior assistant deputy minister in charge of the Water sector of Environment Canada was holding high-level policy and planning meetings in Ottawa on the day of the final meeting. Everyone's attention was on the game which being televised, but the set was in the room adjacent. At long last, the SADM could bear the suspense no longer, and we all flocked next door to see our lads triumph in a very exciting finish. An excellent illustration of managerial leadership!

The second event came toward the end of the year, when I took part in my last Senior Hydrographic management meeting. It was held during the last week in

November, just prior to the annual Grey Cup football game, which was being played at Ivor Wynne Stadium in Hamilton, with the Saskatchewan Roughriders and the Hamilton Tiger Cats prepared to battle for the cup. I had managed to get good seats for all of the hydrographers and some members of our family. Under the circumstances, most of the management team decided to stay over the weekend for the big game. Gerry Ewing, Colin Martin, Mike Bolton, Russ Melanson, and Jake Drake, our branch personnel manager, together with Doreen and myself and our middle son Duncan, all settled in at the big game, the adults well fortified with food and liquor. It was a great game—our team won, but not by much. The weather was perfect—people dropping in by parachute all over the place. The only dull note was having to watch our Prime Minister prance out on to the field to display his lack of talent before the assembled multitude. Doreen invited us back home after the game for great steaming bowls of Irish stew, which soon revived our slightly wilting spirits, and all the out-of-towners headed for home, hopefully reasonably sober. It had been a great day.

The next day winter set in with a vengeance as I set off for a Canada-US planning meeting in Detroit. Another year had gone by, with many changes having occurred. What would 1973 bring?

Chapter Twenty-Eight

Weathering the Bureaucracy – Broadening My Horizons

Harvey Blandford now departed Burlington to take up a position in Ottawa as advisor on marine matters to the Dominion Hydrographer. He had been a sterling assistant to me from the very first day I took over in Central region. I would miss his loyalty, his good advice, and cheerful disposition.

With my replacement as Regional Hydrographer, Adam Kerr, now in place, I had to learn to step back a bit and leave the day-to-day, hands-on management to my successor. It was not an easy thing to do, but I soon found that the bureaucracy was going to keep me busy in its own frustrating way. Meanwhile, I was able to push hard to get our oceanographic activities expanded in the St. Lawrence River, and increasingly so in the High Arctic regions, where we were strongly linked into the Polar Continental Shelf Project. Also, I was able to obtain funding for our Coastal Zone Mapping project, which enabled us to commence work on a major atlas of the Canadian coastal zone in Lakes Ontario, Erie, and Huron. The data collected and compiled for the atlas clearly indicated the cyclical range of water levels over a thirty-year period and highlighted the lemming-like rush to build homes and other structures within the critical coastal zone. Many of these properties were at risk, as they lay well within the cyclical danger line. We planned on completing the Coastal Zone Atlas and publishing it no later than 1974.

Representing Marine Sciences branch at CCIW could be a stressful affair, with a real struggle for power developing within the water sector at both the regional and the headquarters level of involvement. I found myself shuttling between Ottawa and Burlington and becoming ever more frustrated. My left leg swelled up at the knee, and I eventually had to be hospitalized for a few days, suffering from considerable stress. I needed a break, so I headed out to the Pacific coast for meetings with the National Ocean Survey in Seattle and with our own Pacific

Region to discuss and develop plans to exchange key members of our staffs and cooperate on other matters.

While in Seattle I had the opportunity to renew my friendship with Bob Williams, former director of the Detroit office of the US Coast & Geodetic Survey. He was now partially retired but doing some work with the University of Washington on the allocation of funding for ships engaged in scientific studies. My wife was with me, and we were invited out to his home on Tiger Mountain, outside Seattle, where we met his wife and son. After a fine dinner washed down with a suitable wine, he proceeded to show me his trophy room. Earlier in his career, he had spent some time surveying off the coast of Liberia in West Africa. Liberia was a country settled by former slaves from the USA as a goodwill gesture back in the late nineteenth century, and it had an informal tie to the US, which brought about the assignment of a US hydrographic survey vessel to conduct operations leading to the publication of a series of nautical charts of the Liberian coast. During his stay there Bob had amassed a large selection of African wood carvings, many of them quite large and eye catching. He wanted to present me with one, but I managed to decline his offer gracefully. However, he also had a selection of brass ship's instruments, such as bells, clocks, and even a complete compass binnacle. He was absolutely determined that I not leave his home without a gift, so after fighting off his offer of a gigantic ship's bell, I succumbed to his blandishments regarding a solid brass ship's clock. It was a Chelsea clock manufactured in Seattle and had at one time been on board the old survey ship *Rainier*. I was most honoured to accept his gift, even if it did require an overhaul and general adjustment.

Upon re-entering Canada, I declared my gift to Canadian Customs and therefore spent hours with them while they attempted to determine whether this clock was a nautical instrument or just a brass clock. I pointed out that it did not work at present, and guessed for their benefit its possible value. They finally let us go with the clock after we paid duty of eighteen dollars. How they arrived at that figure, heaven only knows—I had my brass clock and was on my way home.

Art Collin had now become Assistant Deputy Minister of the Marine sector within a new grouping of branches that linked us directly with Fisheries. I now found myself reporting directly to him in Ottawa, while establishing formal links with the Great Lakes Biological Laboratory at CCIW. Meeting after meeting took place in Ottawa and across the country as a power struggle developed that would eventually lead to the formation of the department of Fisheries and Oceans, separated from the Department of Environment. All that was yet in the future, but the machinations went on for several years. I had one joy out of these affairs, which often took place in the Maritimes. My wife loves lobster, so I was able to transport live and boiled lobster home to Burlington, where they were consumed rapidly and with abandon.

In late 1973 I was invited to travel to Australia to deliver a paper on hydrographic surveying in Canada to the annual meeting of the Australian Surveyors Association. The invitation was probably related to the fact that I was the incoming president of the Canadian Institute of Surveying. I accepted, and soon I was winging my way across the Pacific Ocean to Tokyo and Hong Kong. As it was a Canadian Pacific flight, I tried to contact Pete Ainsworth, formerly a hydrographer with me on the *Willie J*, who was now the manager of Canadian Pacific operations in Tokyo. However, I ran afoul of the Tokyo telephone system, and had to give up.

Hong Kong in those days was still very much a British crown colony, and still quite Oriental in the way I remembered from my seafaring days. The hotel was comfortable enough, but a big slowly circulating ceiling fan and lots of bamboo screens did remind me of Rangoon or Bombay. After a day's rest, I was on my way to Kuala Lumpur via Bangkok. The flight was Japan Airways, very smooth, very efficient, and very pretty Japanese hostesses. Upon arrival in Kuala Lumpur in the sticky clammy heat of the evening, I was met by my brother Gordon, who conveyed me to the Hilton Hotel in downtown Kuala Lumpur. In those days it was brand new and very comfortable, with air conditioning throughout. It overlooked the race track, which was very popular with the Chinese population who attended the frequent race meets in large numbers. I was astonished at the services of an intimate nature that appeared to be available through room service, but my brother appeared to be nonchalant about such matters, so I kept my own counsel.

Gordon was my younger brother, and he had been in the planting industry in Malaya since 1948, which happened to coincide with the armed uprising of Communist terrorists from their hidden camps in the jungle. He had survived the worst of the troubles, which were largely aimed at the planters and their families and the managers of the tin mines. The country was now an independent nation of the Commonwealth known as Malaysia, and he was the district agent for his company, Barlow Boustead, and manager of Torkington, a large, eight-thousand-acre estate several hours drive north of Kuala Lumpur in the coastal plain around the town of Telek Anson. He had undertaken to introduce me to the surveying community of Malaysia and also to have me visit his plantation and get a feel for his way of life.

I met the president of the Institution Surveyors Malaysia and several other surveyors at a small dinner hosted by my brother in the hotel, and discovered that if I wanted to meet anyone in the hydrographic field, I would have to travel to Singapore, where the Royal Malaysian Navy Hydrographic Office was located. I made arrangements to go there following my visit to Torkington Estate. I did visit the Barlow Boustead office in Kuala Lumpur, which was a revelation! The office was cramped for space, with every space taken up with dusty-looking files and reminding one of a drawing from Dickens' "A Christmas Carol." All that was missing was Bob Cratchit and his tall desk. The company was doing extremely well but was not displaying any signs of its prosperity. Perhaps that was deliberate!

We set off for Torkington at considerable speed. Gordon's driver was an Indian chap who fancied himself as a racing driver. We sped through sleepy villages, frightening dogs and chickens, and upsetting sedentary elders napping in the shade. Our first destination was another plantation, called Riverside, where Gordon assured me that we would be greeted by the manager carrying flagons of cold beer. Unfortunately, Joe Walker was not immediately in evidence, but his wife Muriel appeared from the bungalow and invited us to partake of afternoon tea. She was a very gracious hostess, but our disappointment must have been visible.

It was on then to Torkington, where we arrived at dusk. Gordon's bungalow was spacious but old fashioned, with large ceiling fans throughout, but thank goodness, air conditioning in the bedrooms. When the doors and shutters were closed for the night, a myriad of insects clustered around each exposed light, which indicated a miserable night for anyone caught out in the jungle. Ara, Gordon's housekeeper, took care of our eating needs, which, if I recall, consisted of a very hot curry dish washed down with lots of cold lager.

Gordon was up and away about five in the morning to oversee the various working parties on the plantation and did not return until about nine o'clock, when we settled down to a big Scottish breakfast served out on the balcony in the steaming heat. The bungalow was surrounded by beautiful gardens showing off massive canna lilies and large numbers and varieties of orchid. The fearsome night seemed to have been just a dream. I was then given a grand tour of the estate, meeting the three undermanagers, each with his separate quarters, and then out to the farthest limits of the sprawling plantation. Most of the workers were Tamils, who lived on the estate together with their wives and children. The living quarters included a school, shops, and a medical clinic. There were also places of worship, and even a grog shop. Malays from Kampongs close to the estate also worked in spraying and cutting operations, and they were usually female. My brother seemed to have a good rapport with these women, who chuckled and laughed when he engaged them in conversation.

Figure 67 - With my brother Gordon on his Torkington estate, Malaysia 1973

Torkington's eight thousand acres of flat coastal plain was dissected by numerous small canals leading in from the seacoast. Prior to the 1950s these canals had been the transportation arteries of the plantation trade, which was largely in coconuts and their by-product, copra. In the seventies, the estate was crisscrossed with many roads to enable the product to be transported by road to gathering points for processing, before being shipped out by larger vehicles to Penang or Port Klang. An additional valuable product was cocoa, with its shrubs growing in the shade of the coconut palms. The cocoa pods were processed in a special mill before being transported south for shipment overseas. All in all, the estate was a hive of activity.

On a Sunday, the activity largely ceased, and the planters relaxed and invited their friends on other plantations to a curry luncheon, which started with drinks at noon and finished usually in the early hours of Monday morning. My brother had invited a number of guests, the Walkers and another couple from another plantation, who had two ladies visiting them from Canada. In both cases, the visitors had probably driven forty to fifty miles on jungle roads to sample the curry. It was grand fun, with Ara being helped in the kitchen by a male chef from the nearest village. Much food was consumed, and a great quantity of liquor accompanied it. Even at the height of the emergency, when planters and their families were a prime target for the terrorists, this ritual of Sunday curry get-together was adhered to whenever possible. They were obviously a very stubborn and brave group of people, who worked very hard but knew how to enjoy themselves. Mind you, Gordon almost enlivened the party too much, when he threw a twenty-eight-foot python skin into the kitchen and caused the imported chef to go into hysterics and leave the kitchen, threatening never to return.

The following evening we took the Canadian ladies down to see a coastal village, where we tucked into crab and other delicacies from the sea. It was a very pleasant end to my few days spent in the jungle of the plantation. Gordon returned with me to Kuala Lumpur and saw me off on the plane to Singapore, where I planned to visit the Royal Malaysian Navy base at Woodlands, which lay just across the strait from Johore Bahru. I found Singapore much more westernized than Malaysia, with lots of modern well-appointed hotels. I reported the following morning on board HMM Ship *Woodlands*, which was a strange anomaly, being located in a country other than Malaysia. In the late sixties, Singapore and Malaysia had decided to separate into two separate nation states. The Royal Malaysian Navy was building a new base at Lumut, on the shore of the Malacca Strait, not too far from my brother's location at Torkington. Meanwhile, several ships of the fleet remained at Woodlands, including a small hydrographic division. The division was commanded by a Lt. Comm. James, who was seconded from the Royal Navy. He was not present, but his executive assistant Lieutenant Goh Siew Chong was only too happy to talk to me and show me around his ship and the small hydrographic shore establishment. He was enthusiastic about hydrography and the future of hydrographic surveying in Malaysian waters. I met him later in Malaysia, when he had become a four-ring Captain and had established a well-run organization, with fine vessels and equipment under the Malaysian flag. He was a credit to the profession and to his country.

Before leaving Singapore I visited the site of the old cable station at Keppel Harbour, where I had spent many memorable days on cable ships back in 1946 (see *Mandalay to Norseman*). The area had changed, and I felt very much that I had too. The flight from Singapore to Sydney was uneventful, although the plane seemed to be full of Britishers traveling out to Australia to visit relatives who had emigrated many years previously. It was a Quantas flight with a friendly, if talkative, cabin crew. On arrival I was met by an officer from the Royal Australian Navy hydrographic office, who kindly escorted me to my hotel. It was not far from the hydrographic office, which I visited the following day. It was also close to the famous bridge and all its traffic. I was met by Captain Osborne, the commandant of the hydrographic office, who ensured that I got a thorough tour of the premises. Unfortunately, most of the ships were at sea, but I was able to see many of their field sheets and finished charts. The completed work was similar to our own, but I believe that we were slightly in advance of them in both field and compilation techniques. We agreed to exchange hydrographic information on advances in technique and technology. I was then treated to a fine fish luncheon at an exclusive club nearby, and headed back to my hotel to recover.

The following day I flew to Melbourne, where I was met by an officer of the Australian Surveyors Association and escorted to the hotel selected for their annual conference. It was a medium-sized hotel located out of the city center, but quite comfortable, with a cricket ground in close proximity. Strangely, the official opening ceremony of the conference took place in the Anglican cathedral with much pomp and ceremony. This Australian mix of formality and informality I found perplexing but attractive. There was much to learn about this gigantic island state.

When I had been in Australia last, twenty years previously, beer was the main tipple. Now the country seemed to be turning to wine, and damned good wine it was too. Of course being Australians, they had to defy tradition and, in the heat of a summer barbecue, served the red wine chilled. A very sensible idea! At the barbecue out in the country, I met my first platypus. A strange mixture of genes, it signified to my new Aussie friends the "queerness" or rather the "uniqueness" of everything Australian!

I believe that I was able to establish good relations between our two professional societies, which led later to joint cooperative efforts overseas through our membership in the International Federation of Surveyors and the Commonwealth Association of Surveying and Land Economy. I returned home across the South Pacific Ocean, with a brief stop in Tahiti before heading north toward Vancouver and eventually Burlington. It had been a long but productive trip.

Chapter Twenty-Nine

Pushing the Profession

In addition to my duties as director, Marine Sciences Central, I was finding myself more and more involved in the advancement of my profession. My role as incoming president of the Canadian Institute of Surveying was keeping me busy on both national matters and on the international scene, preparing for the 1974 FIG Congress planned for Washington, D.C. I wanted to ensure that CIS was prominent in all the commissions, but particularly so in Commission Four, the hydrographic commission. I must have driven a few colleagues slightly mad with my demands for perfection.

I was becoming aware of the need for good public relations in getting our message across both at the government level and in matters concerning the profession. We acquired the services of John Hall, an ex-newspaperman, who took on his new duties with enthusiasm. He was an Ulsterman, and very proud of his Irish background. We appeared to have good rapport and so started a relationship that was to last many years. Not only could he write well but he also had a good feel for what I needed to be an effective verbal communicator in getting my message across to the broader public. He also had bags of the Irish blarney that went over well on social occasions. His very attractive wife Sherrin attended these affairs and cut quite a swath among the male attendees.

Early in January 1974 I attended the annual duck dinner hosted by two of our suppliers. They were grand affairs, with lots of drinking and entertainment followed by the actual duck itself. As usual, I consumed more duck than absolutely necessary, and when awakened by an agonizing pain in the middle of the night, naturally assumed that the culprit was the confounded duck. My wife insisted that I be taken into the hospital for examination and treatment, but the emergency team agreed with me that it was probably the duck and sent me home. But it was not the innocent duck—shortly afterward I had to return to hospital, where they finally diagnosed the complaint as kidney stones, a most painful event. The problem was soon fixed, and I was sent home to recuperate under the watchful eye of my better half, who never wanted to hear about duck again.

The usual bureaucratic meetings followed as the struggle for power between Environment and Fisheries became ever more heated, minor empires were created, and well-established entities defended themselves vigorously. After thirty years have gone by, I find that nothing has changed in the federal bureaucracy, as frequent changes in the political structure force drastic changes on the bureaucracy below.

Regionally we were in good shape: the Hydrographic division under the leadership of Adam Kerr was doing well, with operations throughout the region backed up by an active research and development group that were eager to field test their ideas and creations. The chart compilation section was also expanding and making use of the latest technology. The Tidal group had expanded their role to embrace some physical oceanographic studies in both the St. Lawrence River and the High Arctic, mainly through the efforts of Nelson Freeman, who now lived fairly close to us in Aldershot. His wife Torchy was a dear person, and very pregnant with her first son. My eldest daughter described Torchy thus, through the eyes of a teenager: "Dad, she was the littlest–biggest pregnant lady I had ever seen." I hope she does not mind this recollection!

Additionally, hydrographic training was regularly carried out at CCIW, usually at the Intermediate level, as required by our new training schedule. Last but by no means least was our Ship division under Austin Quirk, who manfully made sure that all the many vessels in our fleet were kept fully operational and ready to meet the varied demands of the hydrographers and scientists.

Another activity was the monitoring of surveys contracted out to private industry, which were mostly conducted on the Great Lakes. I hoped that such activities could be expanded in the future to the other regions. There was some resistance to contracting out, with all sorts of silly reasons being given for inaction. It was a trend, started in Canada, which was eventually taken up by other countries. However, the trend faltered in Canada due to reduced funding for charting and a hidebound attitude in management that delivery of the end product must be in the hands of government. It reminds me of the present-day political battle over delivery of our health system.

The biggest event of the year for me was the Canadian Institute of Surveying Annual Conference, which was being held at the University of New Brunswick in Fredericton. I was the incoming president and so had to make many preparations prior to the event, which was scheduled for late June. It was one of the hottest summers on record, and everyone was sweltering in the heat and accompanying humidity. *Bayfield* had been operating in the Gulf of St. Lawrence, so it was decided to have her on display in Fredericton if at all possible during the CIS affair. She was very popular with visiting surveyors and their companions, particularly when she dispensed lots of cold drinks of various descriptions. With the heat, the water level in the river was falling fast, causing the Captain and me some concern that *Bayfield* might not make it over the bar into the deeper water beyond. I released the Captain from his responsibilities a day before the conference ended, and he gratefully pushed off down the river.

Figure 68 - CIS: having fun in Fredericton, June 1974

The conference ended with a bang on the last night, when dinner was followed by a twelve-piece orchestra playing all the tunes we all knew so well. The temperature rose to above one hundred degrees Fahrenheit, causing tuxedo jackets and ties to be discarded, together with items of female clothing, as we frolicked wildly on the large dance room floor. I can still see my wife gyrating around as she was flung around by her companion as they carried out their version of some Israeli dance. It was a grand sweat-soaked finale to my induction as president of CIS.

In the early hours of the morning, Derek Cooper borrowed my car to run down to Saint John to buy some fresh live lobster, and he managed to collide with a large deer that appeared suddenly out of the woods. The poor deer expired, and my vehicle sustained some damage to the hood and driver's door and was partially covered with the blood of the lamented animal. Such was my car when returned to my hotel before dawn. The blood was soon removed, leaving me to wrestle with a creaking door and no repairs in sight until after the weekend. I prayed that no more animals would come crashing out of the forest as we slowly made our way home to Ontario.

The next professional occasion was in September, when I traveled to Washington, D.C., to participate in the Fourteenth International Congress of Surveyors being held at the Washington Hilton Hotel. It was a very big affair, being linked to the annual American Congress of Surveying and Mapping meeting and that of the American Society of Photogrammetry. In addition to the many US delegates, there were hundreds of participants from the member societies of FIG from overseas, and of course there was a large contingent from Canada.

Commission Four, the hydrographic surveying commission of FIG, was headed up by Marc Eyries of France, with Jerry White of the UK as secretary. Dr Fagerholm of Sweden, the designated vice chairman, had taken a UN appointment overseas and was unable to attend. I headed up the Canadian contingent to Commission Four, and also as president of CIS represented Canada in the Permanent Committee deliberations. A hectic schedule ensued.

Commission Four had scheduled twenty-two papers to be presented in six solo sessions and in one joint session with Commissions Five and Six, the other commissions of FIG dealing with allied professionals in engineering and technical applications. Canada had made a good showing, with six papers and a joint effort with the UK and the Netherlands outlining the progress of the Commission Four Working Group on Standards of Competence in Hydrographic Surveying. Other papers were submitted from the UK, Germany, the US, the Netherlands, Ireland, and Japan. It would have been nice to have significant papers from other hydrographically significant countries such as France, Italy, Norway, and Sweden, but that would have to await another day.

The contracting out of hydrographic surveys from hydrographic offices to the private sector was a matter of great interest to all participants, and was well dealt with in papers by Tony Green of Canada and Mile Wright of the UK, both representing the private sector position, and with Roger Morris of the UK hydrographic office presenting the government position. The need for well-thought-out specifications and accuracy in positioning and depth, allied with detailed coverage of the area in question that would eventually appear in the published nautical chart, were all debated at length in the question period that followed the presentation of the papers.

A number of papers followed dealing with the new technology. Neil Anderson of Canada dealt with the use of colour aerial photography in delineating shallow water areas and the impending use of pulsed laser systems to penetrate the water column and enable airborne surveys to conduct accurate hydrographic surveys in clear water to considerable depth—all very promising. This was followed by another Canadian paper by Rick Bryant, Mike Eaton, and Steve Grant on the use of range-range Loran C for precise positioning in conjunction with Sat Nav, presently being used in the Great Lakes but planned for use off the east coast of Canada. The final paper in that session was one by Sleditz of Germany on the Atlas Alpha Doppler Navigator, a positioning system based largely on the utilization of the "Doppler Shift" phenomenon. It sounded very interesting, but seemed to be still undergoing laboratory and seagoing trials.

The next set of papers largely concentrated on the development of automated systems, both for data logging at sea and in designing a suitable automated chart production process. The papers were by Ball of the US, Hope of the UK, Furuya and Evangelatos of Canada, and Bettac of Germany, all describing their own nation's progress in utilizing the possibilities presented by the computer age.

The next paper was delivered by Van Weelde of the Netherlands. He described the Netherlands experience with their new automated system Hydronaut, and the paper was worthy of comment as well because of the author's dramatic and at times hilarious delivery. Van Weelde was a worthy successor to his famous Dutch predecessors, Martin Tromp and Cornelius Tromp, who swept the English fleets from the southern North Sea in the seventeenth century with brooms tied to their mastheads to portray their message of "sweeping clean." Maybe Hydronaut was not a reincarnation of the "broom," but Van Weelde presented the Dutch case very well.

A paper by Colin Weeks of the UK followed describing the operation and evaluation of an automated hydrographic survey system in the Thames estuary as conducted by the Port of London Authority vessel *Mappin*. The system evaluated was Autocarta X, which produced a chart of the area surveyed in fairly short time. There are still headaches, but progress is definitely being made.

I now come to probably the most important paper of the Congress, the Commission Four Working Group report on Standards of Competence for Hydrographic Surveying. This working group was headed up by Alan Ingham of the UK, lead lecturer at the North East London Polytechnic in hydrographic surveying, and former survey officer with Royal Navy Hydrographic Department. He was ably supported by Adam Kerr, Regional Hydrographer

Central, Canadian Hydrographic Service, and Peter Sluiter, Shell International Petroleum, the Netherlands.

The working group was formed after the Wiesbaden Congress to agree upon internationally desirable standards of competence within the profession of surveying at sea, bearing in mind the following interrelated factors that have contributed to the relatively recent expansion of surveying for the offshore industry.

- Contracting out of hydrographic surveys for the production of nautical charts by Government Hydrographic Offices
- Increased draft of commercial shipping
- Increase in coastal and offshore civil engineering projects
- Exploitation of oil & gas resources
- Nature conservancy and control of pollution at sea
- The need for knowledge of the oceans beyond the continental shelves

The WG examined all known existing educational syllabi and methods and organizational structures, before drawing up their own suggested standards at both graduate standard and non-graduate levels. These were thoroughly vetted by consultation with experts in the teaching profession, in the industry, and in government hydrographic offices. The final recommendation of the WG was the establishment of an international board to assess the curriculum of individual teaching institutes in relation to the International Standards of Competence and officially recognize them as custodians of awards given out at the non-graduate or graduate level.

This was a tremendous piece of work carried out by three volunteers, working in addition to their regular duties in the profession. Their work was recognized by the Commission and FIG and by the International Hydrographic Organization, and soon led to the establishment of their recommended International Board for Standards of Hydrographic Competence, which is still in existence in the year 2005, and showing no signs of age or decay. A FIG/IHO category B or A award is very much a prized honour still today. I am very pleased that I played a part in pushing standards of hydrographic competence in FIG and in Canada, and finally was able to play a direct role in bringing counties like Malaysia into the standards of competence orbit.

There were other papers: on unique uses of position fixing by tellurometer by Ridge of the Port of Southampton Authority, Kawakami of Japan on surveying for the Basic Map of the Sea of Japan, and Long of Ireland on offshore surveying in Ireland since Irish independence from Britain in the twenties, which indicated that not much had happened since then, ever the cry of the Irish Victim! My ire was difficult to control! Canada tried to give them advice and assistance later on their problems—we might as well not have bothered!

There was a very good technical paper by my old acquaintance Wentzell of Germany on the Atlas Bottom Mapping Recorder that I had seen in action in the Port of Hamburg back in 1971. It was a good system, well suited to the local requirements of German river waters, but the paper was very much a propaganda piece for the Atlas Company. Two papers on depth measurement followed: Berncastle of the UK on measuring loss of depth in the River Humber estuary, and Roger Cloet on the measuring of the movement of sand waves by side scan sonar in the River Thames estuary, a subject that Roger seems to have made his primary concern.

The final paper was by myself on the demands for hydrographic surveying in the Great Lakes of North America brought about by the rapid growth of uncontrolled industrial and population expansion on the once pristine waters of the largest body of fresh water in the world. The impact of such rapid growth had resulted in major degradation of some parts of the Great Lakes system, and indeed were killing off life as eutrophication took over. Nor was the spontaneous combustion of clogged industrial waste on the Cuyahoga River that flows through Cleveland Harbour in 1949 an isolated event. These warnings were taken seriously in both Canada and the United States, resulting in the implementation of the International Field Year for the Great Lakes (IFYGL). My paper described the IFYGL response, which underlined the immediate need for better hydrographic surveying coverage of the Great Lakes, utilizing electronic position fixing systems that would encompass an entire lake and ensure comparable standards for all the participants in IFYGL.

I have gone into some detail in describing Commission Four affairs at the FIG Washington Congress, in an attempt to emphasize the effect that the FIG connection was having on the world of hydrographic surveying. In a few short years it had become the international forum for the discussion and presentation of new ideas and new technology and techniques, far surpassing the audience available at IHO conferences or even at the narrower, nationally directed conferences put on annually by the Hydrographic Society, the Canadian Hydrographic Association, and a few other national groups. The working groups on Standards of Competence, Positioning Systems, and Data Acquisition Systems were making a major impact in their fields of endeavour. A true marriage of government, industry, and academia in the hydrographic profession world wide would achieve much over the coming years. I was proud to be part of it.

I suppose that I was so committed to this international approach to hydrography as a way of compensating for no longer having a hands-on approach to hydrographic matters in my own region. I had the overall coordinating role and budget control of my region, but did not feel that I could interfere in the policy making or logistical control of the CHS division at CCIW. Indeed, I was sure that my input would not be welcomed by the CHS national management team. Therefore a bloody bureaucrat I had become—immersed in endless meetings that often lead nowhere—but always seeking a wider role for myself and my perceived talents.

During the Washington FIG Congress I talked to a number of American hydrographers who were enthused about the FIG approach and were of great help to the Commission in promoting its activities at the Congress and its aftermath. Of particular note were Bob Munson, the US delegate to Commission Four, Jerry Umbach, who had previously presented a paper at a Canadian hydrographic conference, and Jack Wallace, who presented a paper on depth and the possible approaches to automation. These three men opened up doors of friendship for me with the professionals in charge of the National Ocean Survey in the United States, for which I am ever grateful. Bob Munson became a very close friend over the years, together with his wife Loretta. Jerry Umbach unfortunately died far too young, but Jack Wallace is still active in the profession as the executive director of The Hydrographic Society of America (THSOA).

Chapter Thirty

Fifty Years – How Could That Be?

1975 started out to be a busy year, with travel to Ottawa being high on my agenda. The scheming to create a new department of Fisheries and Oceans was coming closer to fruition daily. However, late in January when back in Burlington, I suddenly came face to face with a much more personal matter. My chief of Administration and Finance, Art Appleby, had invited me to a party at his home. Upon arrival, I was astounded to find out that the party was being held in my honour and to celebrate my fiftieth birthday. It had all been arranged, with everyone including my own wife in the know except me. It was a nice affair and I was duly grateful to all concerned, but thunderstruck to realize that half a century had gone by and that I had scarcely noticed. Indeed, I had tended to view my birthday as of little importance to anyone except myself. Now out of nowhere, I was aware of my mortality and not too sure what to do about it.

Figure 69 - Fifty and suddenly aware of it! 1975

There were, however, more immediate pressing matters to consider. The CHA and the CHS were mounting the annual hydrographic conference in Hamilton, and a number of foreign and overseas visitors would be attending. It was a great pleasure for my wife and me to welcome many of those attending the conference from outside Canada. The US contingent was headed up by Rear Admiral Allan Powell of the National Ocean Survey, with Captain Bob Munson assisting. It was a useful experience to renew and enhance the close relationship and rapport between Canada and the US on hydrographic matters.

Away from the job in Burlington, I was active in the Rotary Club of Burlington, and was becoming active in both the Marine Club of Toronto and the Great Lakes branch of the Master Mariners of Canada. Even so, I managed to keep linked to my family, with much help from my dear wife. I even toyed with the political scene but never got overly involved, being a civil servant. If I had been of independent financial means, I would have loved to become a more active participant in the development of Conservative Party policy. My views had become steadily more right wing as the years had passed, and the Trudeau Liberals were pushing me there faster by the day. Perhaps it is just as well that I did not become a politician—it can be a very dirty business and is not for the faint of heart. But I knew that I could express my ideas well on paper and in debate, and that I might have contributed something worthwhile to the political scene. Even now, in my eighty-first year, I still have faint regrets that I did not have a go!

In the late spring I traveled to Stockholm en route to the FIG Permanent Committee meeting in Helsinki, Sweden. It was my first visit to the Baltic, and I was most impressed with the atmosphere and the people that I encountered. In Sweden I was well looked after by the Swedish Hydrographer, Hallbjorner, and his very efficient staff. I visited the hydrographic establishment south of Stockholm and viewed their progress in automation and their development of the fast "Stick" sounding launch approach of surveying an area with several small, fast, sounding launches in close proximity to each other, thereby obtaining complete bottom coverage. It was perhaps a method not suitable for all areas, but definitely one that showed promise for some high-priority areas demanding better sounding coverage. Hallbjorner then kindly took time off to give me a personal tour of the *Wasa*, the famous Swedish warship that collapsed and sank in Stockholm Harbour three hundred years ago. It had been raised from the bottom and carefully treated so that it could be viewed by the general public. I found Stockholm Harbour to be a most attractive place, and as my hotel room overlooked it, was able to soak up the scenes at my leisure. I was also entertained in my room by the delivery of my breakfast each morning by a beautiful young Swedish maiden. A grand interlude.

This interlude was followed by a journey across the Baltic Sea to Helsinki on one of these very large ferries that run between Sweden and Finland and other countries fronting the Baltic and its bays and gulfs. When boarding the vessel in the late afternoon, I was directed to a cabin located in the bowels of the ship, right up against the engine room bulkhead and without any opening other than the door. It also had two bunks, indicating that I would probably be sharing the space. Whether that would be sharing with a male or a female I never did find out, as I just had to get a cabin with a view of the sea. Accordingly, I pestered the purser, who after a period of impressing me with the great difficulty of finding such a cabin, produced an excellent one just abaft of the bridge. I settled in and quaffed an ale or two while contemplating the scenery, both of the landscape and the more human variety. As we wended our way through the islands on our way into the Baltic, I decided to tour the ship and scan its attractions. The ship was absolutely packed with Swedes on their way to a short holiday in Finland, all seemingly determined to drink all the duty-free liquor on board before we reached our destination. The dining cabin was well appointed, served good food, and had a full dance orchestra, which played music that made you just ache to dance. In no time at all, I found myself in the company of a convivial bunch and was able to attempt to show off my talents on the dance floor. This continued even when we hit some stormy weather and sort of slithered around the floor with great caution. After midnight, our party adjourned to the Disco, which was located away down below, not far from my initial cabin space. We continued sporting ourselves until about three o'clock in the morning, when the Disco shut down and we were proceeding up the Gulf of Finland. I parted from my convivial Swedish friends swearing undying regard for one another, and made my weary way up to my well-appointed cabin behind the bridge. It was a sad and shaky hydrographer who rolled off the ship's gangway in Helsinki.

I was greeted by Hugh O'Donnell and other members of the Canadian FIG contingent, who wanted me to convene a planning meeting immediately to prepare for the following week's activities and to meet with the Canadian Ambassador regarding a proposed cocktail party for FIG officials. I desperately needed rest but was thrown into a hectic flurry of activity. Never again, I swore, would I overindulge on a Baltic crossing!

The Permanent Committee meeting in Helsinki was a great affair, with the Finns doing their very best to make us feel welcome. Helsinki's historic buildings were surrounded by lakes and parks, with the ancient harbour dominating its approaches from seaward. Our hotel was ultra modern, as were many of the buildings out beyond the city core. Finland had been in bad shape in the aftermath of World War II, but was gradually building up her economy, while fending off the political pressure of the Russian Bear next door, who had made Finland pay dearly for being on the losing side. The Finns liked to trade with the West and exchange ideas freely but had to continue to keep a close eye on their next-door neighbour. It was, after all 1975, and another fifteen years would have to pass before the Soviet Union was no longer a threat.

The Hydrographic Commission held several meetings, discussing the work of the various working groups and the progress of the commission initiative to determine the standards of competence for hydrographic surveying. Both our chairman, Marc Eyries of France, and our secretary, Jerry White of the UK, were present, and welcomed me in my new capacity as vice chairman of the commission. In addition to planning for the next FIG Congress to be held in Stockholm in 1977, we drew up plans for the FIG Permanent Committee meeting planned for Paris in 1976—the centennial of the formation of FIG.

In the main Permanent Committee sessions, the Canadian Institute of Surveying made its first proposal for consideration as the host of a future Congress, which finally became reality in 1986 when we held the Congress in Toronto. Our main competition were the Swiss and the Bulgarians, and a friendly rivalry developed over the years, particularly between the Communist Bulgars and ourselves.

We departed Helsinki well pleased with ourselves, with our CIS group preparing themselves for the CIS Annual Meeting, which was planned for Winnipeg in late June. It would mark the end of my year as president of that august body, where my major contribution had been the promotion of the institute on the world surveying stage.

During my year as president of the Canadian Institute of Surveying, I had been greatly helped by Jim O'Neill, the Executive Manager of our organization. He was a former surveyor with the Surveys and Mapping Division of the federal department of Energy Mines and Resources, and knew everyone in the surveying and mapping business. As I was the first hydrographic surveyor to head up CIS, Jim's knowledge of the national and provincial surveying scene was of considerable assistance to me in making sure that all the various assorted parts of the profession were involved in our decision-making and policy development. He also ensured that I was well briefed for the Winnipeg meeting.

Members of the CIS board such as Dave Usher, Charlie Weir, Hugh O'Donnell, and John Pope also provided much valued support and friendship during my period of office. I had very mixed feelings as I handed over the reins of office to my successor—some pride in our achievements and, while glad to hand over the burden of office, some regret that I could not continue to influence the day-to-day direction of the institute.

The business of the CIS Annual Meeting went without a hitch, while we braved the warm weather in Winnipeg, which brought out a plague of mosquitoes. It also coincided with a possible strike by Air Canada staff that threatened to torpedo our arrangements, but it was averted at the last minute. This travel upset was of particular concern to me, as I had invited a number of people from foreign surveying and mapping organizations to attend and participate in our deliberations. Among them was Robert Steel, the executive director of the Royal Institute of Chartered Surveyors of the UK and also the Commonwealth Association of Surveying and Land Economy, an organization in which CIS was interested in perhaps pursuing membership. Together with all the other delegates, he joined us on the final evening event—a cruise on the Red River. It was a most romantic affair, cruising along the river in the setting sun and indulging in the usual eating and drinking and dancing, while fending off the odd venturesome mosquito. A really Canadian event.

Turning back to the more mundane matters of my governmental responsibilities, I immersed myself in administrative matters prior to traveling to Quebec City and the Tadoussac region to take part in oceanographic studies being conducted from CSS *Bayfield* by members of Nelson Freeman's staff. Our daughter Ellen accompanied me on this trip. She was an attractive young lady, so I knew that she would be welcomed by the *Bayfield* gang. I would, however, be required to keep a fatherly close eye on her and any of the more ardent members of our crew. She was good company for me on the drive to Quebec City, but managed to create problems upon arrival at our hotel by inadvertently locking and then losing the keys to the car. Daughter was distraught and father was fuming, but all was fixed by the following morning.

We boarded *Bayfield* and steamed away downriver to the vicinity of La Malbaie, where I was able to watch the ship in action as current meters and other underwater probes were deployed to collect oceanographic information. The ship seemed well suited to this type of work, and the scientists and technicians were enthusiastic about their work, while the Captain and crew were thoroughly professional. One of the technicians, Sam Baird, was from my home town, Greenock, and he took that opportunity to belabour me directly about the project's urgent need for additional funding. I could scarcely disagree with him, for I was in full support of the project, but had already had to divert some funding from other divisions to keep his operations afloat.

We were unable to stay on board as there were no free bunks, so Ellen and I found a small motel near Point au Pic where we overnighted. It was very close to the magnificent Manoir Richelieu, which was way beyond my budget limit. We resumed operations on board *Bayfield* the following day and then were transported back to Quebec City. We proceeded home to Burlington the following day. Daughter was happy with her experience, although frustrated by her inability to make herself understood in French, particularly in ordering a meal in Point au Pic. Father was pleased with his visit to the project and happy that daughter was happy!

My next venture was unusual, as it took place in Disney World, Florida. There was a conference scheduled to be held in the Contemporary Hotel inside Disney World featuring the uses of satellites in space, with particular emphasis on coastal and oceanic charting and interpretation. Being so close to Cape Canaveral, we were given a tour of the establishment and were treated to a presentation on the uses of high-resolution satellite photography in Law of the Sea disputes on territorial boundaries. The presenter was none other than Werner Von Braun, the German rocket scientist who was leading the US scientific team at the Cape. A major matter of dispute at the time was the so called "cod war" between the UK and Iceland, which threatened to break out into something more serious as gunboats from both countries forcefully protected their fishing fleets from interference in plying their trade in the disputed zone. Werner Von Braun was able to illustrate, by interpretation of the aerial photography, where the actual shoals of cod were located at the time. As both Iceland and the UK were allies of the US, it was thought prudent to keep this information secret. There were many additional presentations, particularly on coastal interpretation, that were most interesting and informative. The techniques were particularly relevant to our work on coastal zone mapping on the Great Lakes.

There were a number of National Ocean Survey officers in attendance at Disney World, among them Jerry Umbach and Wes Hall, who made sure that my hours of relaxation were fully enjoyed. In addition to quaffing many an ale at the Contemporary Hotel, we roamed across Central Florida seeking a decent belly dancer, but reluctantly had to finally accept that Central Florida was bereft of the breed. I then headed for Smyrna Beach where one of my staff, Ed Thompson, was on exchange assignment with NOS on board a large hydrographic launch. The assignment appeared on the surface to be going well, but my reception was cool and very formal, not to usual NOS standards. A female ensign was practically rude, and the NOS officer in charge appeared indifferent as to the success of the exchange of officers between national offices. Thompson was unable to shed any light on the matter, so I was left feeling that perhaps our training exchange was not going as well as I had expected.

I relaxed for a couple of days at Daytona Beach, now almost deserted in the off season, and then flew to Norfolk, Virginia, for meetings at the NOS base there. Their base was well laid out and fitted to serve their Atlantic fleet of ships and launches and to compile the hydrographic and oceanographic data received from the field operations. I was impressed, and even more so, when Jerry Umbach took me to see the best belly dancer in Norfolk. I returned home ready to face anything that Ottawa threw at me.

Before I could settle down properly, I was off to the UK to attend a CASLE meeting and to make an official visit to the Royal Navy Hydrographic Office in Taunton. I was an observer for CIS at this CASLE planning meeting, being held at the Royal Institute of Surveying building, which is located close to Parliament Square and Westminster Abbey. The mix of history and pomp and circumstance played well to my psyche and I became an enthusiastic supporter of a Canadian initiative to join CASLE at the earliest opportunity.

CASLE—the Commonwealth Association of Surveying and Land Economy—was founded 1969 in London, England, and has member associations in most member countries of the British Commonwealth. It links the three separate but related disciplines of Land Economy, Land Surveying, and Quantity Surveying, mirroring the structure of the Royal Institution of Chartered Surveyors (RICS). Hydrographic surveying fitted under the Land Surveying banner. CASLE's stated objectives were the following:

- To maintain and strengthen professional links between Commonwealth countries.
- To foster the establishment of professional societies in countries where none presently exists.
- To foster appropriate standards of education and opportunity training.
- To disseminate information and foster development of research and technical information.
- To facilitate personal contacts within the profession.

It certainly appeared to be a worthy organization deserving of Canadian Institute of Surveying support.

My visit to the establishment at Taunton was fairly routine, but I was given a personal tour by David Haslam, who was now the RN Hydrographer. I was most impressed by the cartographic department, where they kept an inventory of the many thousands of nautical charts that represented worldwide coverage. Their steps taken toward automation seemed, however, less certain and less focused than our own. I ended up the UK visit by attending to family matters. There were the Cullens in London, Frank, Margaret, and their daughter Maria, who were related to me through my wife. Finally, north of the border, I was able to contact my Aunt Bessie and keep the Scottish connection strong.

In November I traveled to Halifax with my wife to participate in a Canadian Institute of Surveying meeting hosted by the Association of Nova Scotia Land Surveyors. It was a very special occasion, where normally I would have been expected to address the attendees. Instead, 1975 being designated as International Woman's Year, my good lady had been invited to address the gathering. She worked hard preparing her speech and was able to wow her audience with her story of being the wife of a surveyor. It was funny, it was pointed, and it thoroughly won over all those assembled at the luncheon. It deserves inclusion in this narrative. It was also printed in the Canadian Surveyor and earned her a small financial award. Indeed the world was changing!

THE WIVES' POINT OF VIEW

Presented at the CIS Convention, Halifax, Nov 1975

Madam Chairperson?
Madam Chairman, Ladies and Gentlemen:

I should begin by saying how happy I am to be here speaking to you; I won't, because I won't be happy until it is all over.

We all know that this is International Women's Year. I am sure most of the wives will have informed their husbands of this by now, in one way or another. So when Beth Thompson wrote to me in the spring to ask me to be the guest speaker for today's luncheon, I thought that I had better do my bit for Woman's Lib and all that, by accepting the challenge. On checking with my family, my daughter's reaction was "Oh, for heaven's sake, mother, go ahead and do something on your own first without asking Dad." So I jotted down a few ideas on the topic suggested and when Tom phoned from Europe to see what was new at home, I casually mentioned the fact that I had been invited to Halifax in November to speak at the Halifax Branch CIS Luncheon. He was only half listening as usual and said, "Well, now that shouldn't be any problem as I'll be there anyway...Yes, I'll be only too happy to be the guest speaker." " No," I said, "not you...me"... He was silent for a minute or two and then just laughed and laughed ... "Oh, no" he said, "Not you! What are you going to do?" "Go ahead with it," I said. "Good for you," he said, and kept on chuckling until he hung up.

This gave me all the confidence I needed – to refuse. However, when we went to the Annual CIS Convention in Fredericton in June, I changed my mind. Did you see that head table? Was there a woman at the mile-wide table? Did anyone acknowledge the president's wife...with all the trials and tribulations she has had to put up with during his term of office? Was she thanked or presented with a bouquet? Oh, no; but mention was made that the president did have a wife. I wonder if things will be any different at the next convention. I am not one to throw out a few hints but ...

Well, it was obvious that some changes would be made, and I congratulate the ladies of the Halifax Branch of CIS for taking over from these male chauvinists today. We should also salute Bill Thompson for his bravery in going along with it. He may yet be drummed out of the club.

Now, it was time to do some research into the art of speech making. The men are so practiced at it and do such a good job. I felt that I had to do something in contrast. In spite of the fact that it is International Woman's Year, we don't want to compete with our husbands, or do we? There are a few rules that should be followed when making a speech.

Rule No. 1 – was broken at the start...never begin with the word ' I' and use it as little as possible ... doesn't make too much sense ... we all love to talk about ourselves, and what wife would waste an opportunity of making her husband sit and listen to her for a change.

Rule No. 2 – keep in mind the nature of your audience.... Avoid going over their heads by using technical terms...that's easy enough...if there's anything you men can't understand, then I'm sure that your wives will be only too pleased to explain – in detail.

Rule No. 3 – use quotations – Tom's favourite quotation from Kipling – "to a rag , a bone, a hank of hair , etc.," wouldn't do at all. More appropriate is ... "Behind every successful man is a woman with nothing to wear."

Rule No. 4 – is the only that I'll go along with... Keep your speech short.

Rules are only guidelines... they're not too logical anyway.

Oh, yes, the title of my speech is "The Role of the Wife of a CIS Member," First of all I'd better explain what a CIS member is. He's the usual business-type male chauvinist, doing a good job in the survey fields. None better, I might add, when he surveyed for a wife. Rather an absent-minded bunch where their wives are concerned, but now and then they find time to spend with us. They don't mind bringing us along on conventions, sometimes joining in the ladies' programs. For instance, a few weeks ago in Worcester, Massachusetts, one or two husbands took time off from a most important and vitally interesting morning session, to spend an hour or two with their wives at a poolside fashion show. You can imagine how bored they were watching these gorgeous models, but they stayed to the bitter end; the finale was a bikini parade.

To survive as the wife of a CIS member, one needs to have a keen sense of humour. On the whole, our lives are pretty humdrum, but our husbands manage to put a little excitement into it. They come home either down in the doldrums, because everything has gone wrong, or they're on top of the world because everything has gone right. Either way, they create difficulties. When they are down, they complain about everything, particularly the lunch they had that day...caterers never use their imagination... or they go on to describe the lunch in detail. You've guessed it ... that's the meal you planned for supper. You cope with this by making sure that his scotch and soda is extra strong, fancy up some leftover stew, and serve it with one of his favourite wines. If that doesn't cheer him up, warn the kids to keep out of the way and take yourself off to the bingo, where you may have better luck.

When things have gone well with him, he'll come in full of vim, vigour, and vitality saying let's celebrate and go out to dinner for a change tonight. Have you noticed that this will be one of your worst days? You had car trouble and had to cancel your hairdressing appointment that morning, and your favourite show that you had planned to watch is going to be on TV, but, being a dutiful wife, you go ahead and get all dolled up. Meanwhile, he takes a short nap and, when you're all psyched up and ready to go, you'll find that short nap has turned into his night's sleep. No wonder you can hardly keep the smile off your face at times when he says he is sorry he's going to be out of town all next week. Now I'm not saying that all CIS members are like this. Maybe mine is the worst one... guess that's why they made him president.

Better not say too much more or I'll be out of a job. After all, being a wife is a full-time job...and an important one. Where else would you get such good pay? Fringe benefits – like a pat on the bum now and again; production bonuses – like five children; travel – to the stores, school, laundry, football practices, ballet classes, etc., and the excitement and glamour of business trips with your husband.

Mind you we do get time to do our own thing, like bowling, bridge, coffee parties, and a luncheon now and again. So believe me, we're not complaining – we wouldn't dare.

So being a wife is a full-time job, being a mother is more demanding, but being the wife of a CIS member is having to be superwoman.

Now I've had my fun, I'll relax and enjoy myself.

END OF SPEECH

It was quite a performance and very well received. I fully expected her to follow this fine effort with other significant pronouncements, but she never showed the slightest interest in going public again, and saves her most scathing comments for this author's ears only. There were, however, several interesting observations from members of the surveying community, such as "You couldn't have written that— you had help from your husband!" Absolutely untrue and insulting to boot. It was all her own work, and she deserves much credit. Then there was the esteemed academic who was most upset that she would receive an award for her speech, while he, after having several scientific articles published in the Canadian Surveyor, had never received such recognition. That he would say these things directly to her in public was an indication of both his fury and his pettiness. When I was told of the incident, I was astonished and then angry, but was persuaded by the wife of a CIS member to ignore the silly man.

Her performance was such a success that I was persuaded to take her with me to Trinidad several weeks later when I was invited to attend a CASLE regional meeting dedicated to Caribbean affairs. We arrived in our hotel after dark and ready to rest after a lengthy flight from Canada, but were met by Bob Steel, who persuaded us to join a small dinner party being held to honour the president of CASLE, who was visiting from the UK. So, two slightly befuddled characters from Canada joined the select group in the private dining room of the hotel. It was a kaleidoscope of colour, with black, brown, and white faces, while the women sported a mixture of bright and spectacular dresses and the men struggled in the tropical heat inside their darker suits and tight shirt collars and ties.

We really were made very welcome by most of those present, but it took us a little while to get our bearings and discover who was who. In addition to the president and his wife, there were CASLE representatives from most of the Commonwealth nations of the Caribbean area, such as Trinidad, of course, as the host nation, and Barbados, Jamaica, Grenada, and St. Lucia, just to mention a few.

I soon found myself in conversation with Sir Oliver Chesterton, the president of CASLE, who gave me a broad-brush history of CASLE since its inception in 1969. He explained that Aubrey Barker of Guyana was to have been the first president of CASLE, but he died suddenly, and the 1969 general assembly in Guyana had elected him in Aubrey Barker's place. He was looking for a replacement as soon as possible and indicated that John Bloomfield of Jamaica, who was present, was his likely successor. Meanwhile, one of my table companions, a pert, pretty lady from Trinidad, was explaining to me in great detail why she returned to Trinidad after a number of years in Britain. I gathered that in her view, racism was at the heart of her decision to return. Another lady, from Tobago, wanted to ensure that I visited Tobago before returning to Canada. Meanwhile, my wife just soaked up the warm tropical atmosphere and enjoyed her first visit to the Caribbean.

There was an official opening where I was supposed to produce our High Commissioner. Unfortunately, he was unable to attend but sent a substitute official in his place, who did not identify himself and therefore left me looking silly in the foyer. He turned out to be a hippy-looking fellow who sidled into a rear row of seats. Not exactly one of Canada's better efforts. Every other country was well represented, including Venezuela and the United States.

The sessions on surveying education and codes of good practice and conduct were most interesting in the main, although I did envy my wife, who was off on ladies' jaunts every day before joining the rest of us at official functions in the evenings. I met there a number of surveyors with whom I have retained friendship for many years: John Bloomfield and George McFarlane of Jamaica, Randolph Choo Shee Nam of Guyana, and Mike Griffiths of Barbados. During these evening events, we were personally escorted by a young Trinidadian couple; he was a quantity surveyor and she managed a medical center. They held a gigantic party in their home up in the hills one evening with a full Trinidad steel band in attendance. It was such a great affair and so washed down with lots of rum that we were unable to go on a scheduled trip to Tobago the following day. Never mind, we have since been in Tobago several times and love it.

And so ended a tumultuous year when I celebrated my half century.

Chapter Thirty-One

The Pace Quickens

The year started off slowly, but soon speeded up as the Fisheries and Marine sector became autonomous within the former department of Environment. The pressure had been building up for some time, and it was a relief to have it all settled, even if the Fisheries sector appeared to have undue influence over our mandate regarding marine sciences and surveys. Art Collin became Assistant Deputy Minister for Atmospheric Environment, and I no longer had direct access to the man whom I regarded as my mentor in the departmental corridors of power. This was eventually to affect my linkage to Ottawa and perhaps my prospects for further advancement

We no longer had to interface with our regional counterparts in Wildlife and Weather Analyses and Prediction, plus other assorted parts of the environmental envelope, except in the continued provision of vessels to the Inland Waters Directorate as agreed to between the two parties. So no more meetings in Calgary or Saskatoon, but lots more in every part of Canada where fisheries and marine sciences interface. Far too many meetings for no apparent reason other than because it had always been so, or to prop up the ego of a senior bureaucrat. However, it did give me many useful visits to the Maritimes, where I was able to continue to indulge in procuring lobster for my appreciative family.

At home, in addition to my involvement with Rotary, I had become active in the Company of Master Mariners and in the Marine Club of Toronto. Indeed, I was now vice president of the Marine Club and in charge of arrangements for speakers at our annual dinner. I had the job of looking after George Steinbrenner, the owner of the New York Yankees baseball team and the scion of a well-known shipping and ship construction company based on the Great Lakes. He was quite a handful, having very strong likes and dislikes, which he made known in a rather peremptory manner. However, he gave a good speech, which was well received by his audience of sixteen hundred well-fed and slightly inebriated members and guests assembled in the Royal York Hotel. Just for the record, a gentleman by the name of Paul Martin sat next to me at the head table. He was the president of Canada Steamship Lines and, as you all know, later became prime minister.

Our regional oceanographic effort was still advancing under the direction of Nelson Freeman. In addition to our work in the St. Lawrence River, our work in the High Arctic with the Polar Continental Shelf Project proceeded apace. Not even forays by large polar bears into their tented living areas out on the polar ice could discourage our field staff engaged in their scientific studies. These physical oceanographic studies in the High Arctic were unique, and neither Patricia Bay nor the Bedford Institute could mount such a project at this particular time. Our gang were rightly proud of themselves.

There were two trips overseas in 1976, the first one to Kingston, Jamaica, to take part in CASLE Atlantic regional meeting. After the success of the Trinidad affair, I brought my wife along to enjoy the ambience. It was a time of political unrest, and we were swept up in the excitement shortly after we arrived at the Pegasus Hotel in New Kingston. As we left the hotel for an after-breakfast stroll, large numbers of people started gathering in New Kingston, and we found ourselves trapped in the doorway of a shop as they started assembling before marching down to the city centre. Luckily, they were in an excellent mood, and joked with us as they plastered posters over every glass surface that they could find—shop windows, car windscreens, all were fair game—knowing full well that that these posters would stick hard to the glass surfaces as the temperature inevitably rose. Then they marched off carrying banners and shouting slogans as they headed into the main city. Two very relieved Canadians snuck back to the relative quiet and security of their hotel. Later that day our marchers clashed with their opponents, and many people were hurt. The meeting was a bit of an anti-climax after that first morning wake-up call, but I was able to renew acquaintance with many surveyors that I had met in Trinidad. George MacFarlane and Dr. Sangster were particularly friendly as were, of course, John Bloomfield the new president of CASLE, and Robert Steel, the secretary. I also got to know the other members of the Atlantic region executive, Victor Hart from Trinidad and Lionel Patterson from Jamaica. The mixture of land surveyors, quantity surveyors, and land economy surveyors was still proving a bit of a puzzle to me, a poor deluded hydrographic surveyor. But there was good will on all sides, so we set to, to develop understanding among us and thereby convey our expectations to the appropriate levels of government.

Contact was made with the hydrographic unit operated by the Land Survey division. They were quite enthusiastic but lacked training, properly maintained launches, and equipment. I agreed to have additional meetings with Harry

Armstrong, the director of Surveys, to explore ways and means for Canada to assist Jamaica in meeting its hydrographic goals.

A speculative comparison between Jamaicans and Trinidadians appeared to show that Jamaicans tended to be more serious and even aggressive than the fun-loving Trinidadians that I knew. Each place seemed to see itself as the logical leader of the islands of the former British Caribbean, and thereby rubbed the other the wrong way. Each had great expectations that eventually proved to be overinflated. I trod my way carefully between the two camps.

Kingston was a city of some history, dating back to the battles for possession of Jamaica between England and Spain. It had been built after the destruction of Port Royal by earthquake in the seventeenth century. The old city, built down on the waterfront and climbing the hills, was gradually deteriorating, while New Kingston, further up in the hills, flourished with newer buildings and shops supported by enclaves of the middle and professional classes. In the lower city, there were large areas of run-down property filled with supporters of the political parties who appeared to loathe one another with some venom. These factions and the general air of poverty everywhere to be seen gave warning of great danger and trouble to come. Meanwhile, we indulged ourselves at the Pegasus in air-conditioned comfort and sported ourselves around the pool.

Our Jamaican hosts entertained us royally each day with luncheons and dinner affairs. Canada even managed to host a small cocktail party, which was greatly appreciated by all present.

It was time to get back to normal living, so we soon headed back home.

The other trip overseas was all the way to the FIG Permanent Committee meeting in Nigeria. I had hydrographic business meetings in the UK, and of course family matters to attend to, before joining the expedition to Ibadan in Nigeria. In London I met up with Hugh O'Donnell of CIS and traveled out to Gatwick Airport, south of London, for the flight to West Africa. I was driving a car and returning it to the airport, but managed to get lost somewhere around Croydon before finally delivering car and passengers safely to Gatwick. We were flying Nigerian Airways to Lagos, and on board met up with several other FIG delegates, including Terry Sudway, an acquaintance of ours from Ireland. The service on the plane was not very good, but we enjoyed ourselves yarning and consuming a fair amount of cold lager. We were met at the airport by Nigerian surveyors, who deposited us in a run-down hotel for a few hours before the bus left to take us up the highway to Ife and then on to Ibadan. The hotel was filled with wild-eyed characters, who threatened us and generally created a most unnerving atmosphere. We were damned glad to be back on the road again. It was a recently built highway in good condition, but was lined with the burnt-out wrecks of vehicles on either side, often pushed away into the jungle. Needless to say, our vehicle was not air conditioned, and frequent pit stops were necessary to acquire cold drinks from little shops in small villages, and to pump ship as required.

After seven or so hours on the road, we arrived in Ibadan. Our hotel looked fairly modern but lacked many amenities that we soft westerners were used to, such as proper air conditioning to take care of the damp, muggy tropical air and to keep out the hordes of winged insects that seemed to be everywhere. The food too was almost impossible, as we soon found out. Even signing into the hotel was memorable, as over the desk was a huge picture of some Nigerian general who had just won the latest coup, glaring at us for daring to bring our quasi-colonial ideas into his land. We found out that we were best served for food if we stuck to local fruit, eggs in any creation, and dainty watercress sandwiches for afternoon tea. Security was almost non-existent, and touts and whores roamed the hallways at night. All in all, not much of a monument to the progress of post-colonial Africa.

Ibadan was a fairly large city, situated in Oyo State, north of Lagos. It was a centre of learning, with a large university and other educational institutes. That is why it had been selected as host city for the FIG Permanent Committee meeting. The meetings were held mostly in the university, and tours were arranged for delegates to visit various surveying training establishments in both Ibadan and in the other university city, Ife. They were to be commended for their progress from colonial status to independent nation, but something was missing in their approach to the modern world. Perhaps it was because of the terrible civil war that had raged in Nigeria until fairly recently, or perhaps the many military coups that had unsettled the country. Whatever it was, the country lacked focus, and that could be detected also in the places of learning so painstakingly erected by the former colonial power. Maybe it was also the corrupting influence of oil wealth, or the sharp divide between the Moslem north and the Christian south. Whatever it was, this large country at the heart of West Africa was acting like a giant that had lost its way.

The Nigerian surveyors had worked hard to make the meeting a success, in particular Chief Coker, the leader of the Nigerian delegation. However, the lack of discipline shown by the population generally made it a difficult undertaking. A grand reception at the former residence of the colonial governor became chaotic as the serving staff enthusiastically joined the official guests in demolishing the liquor and food. It was like a scene out of the French Revolution. Yet, the local Rotary club could put on a well-run meeting that all we Rotarians from overseas could enjoy and relate to. Finally, we found by chance a Lebanese restaurant, shrouded in darkness, that provided excellent French cuisine in this strange land of contrasts.

There were two items of main concern determined by the permanent committee. The first was an illustration of international politics at work. The South African Surveyors Association invited FIG to have a Permanent Committee meeting in that country or to host a Congress, which was their right according to FIG statute. The reaction to this proposal by some delegates and the ensuing heated discussion indicated the extent of politicization of what should have been a purely professional and technical international society. Both South Africa and Rhodesia were refused visas by the Nigerian government. All of this display of anger was aimed at the racial policies of the government of South Africa of the day.

The second item of heated discussion centred around a vote held to determine the host country for the 1983 Congress. Seventeen countries were present to vote on the issue. Professor Pevsky of Bulgaria gave an emotional but stirring presentation—relying on seniority, his honorary member status, and the beauties of Bulgaria to get his message across. Solid support from the Bulgarian association and government was promised. He followed up with a film, which mainly showed Bulgarian pulchritude cavorting on the beaches of the Black Sea.

Our presentation was factual, citing the high level of support we would receive from the leading six organizations in the surveying and appraising professions, our recognized expertise in survey techniques and technology, and the success attending our holding of other international congresses, such as ICA and ISP in 1972. It was followed by a number of slides showing views of the city of Toronto.

We lost the 1983 Congress to Bulgaria by a vote of 9 to 8. Bulgaria was strongly endorsed by all the Iron Curtain countries, and our only verbal support on the floor came from Ireland. The US may well have voted for Bulgaria, leaving us supported by Ireland, Australia, Denmark, Germany, France, Switzerland, and Sweden. I believe the key to our failure in Ibadan was the non-appearance of the British delegation, and to a lesser extent the Netherlands, Belgium, and Norway. I conveyed Canada's congratulations to Professor Pevsky, but suggested that the 9/8 vote indicated a strong level of support for Canada and that a decision be taken as soon as possible regarding the venue of the 1986 Congress. I requested that a decision on this matter be made at the 1977 FIG Stockholm Congress.

I was not unhappy to leave Ibadan, and departed with several German surveyors who had hired a small vehicle and driver. We drove to the hell of Lagos in relative comfort. Getting to the airport was difficult, and then getting aboard one's aircraft was even more so. If I had been on a European airline, my situation would have been easier, but holding a Nigerian Airlines ticket subjected me to all sorts of harassment before I was safely on board my aircraft. Bribery was necessary to ensure my passage, as I had to pay out to agents, customs, police, etc. I believe it cost me about thirty-five dollars US and worth every penny. We stopped briefly in Kano on our way out of Nigeria, and then Amsterdam and finally London. I felt like getting down on my knees and kissing the tarmac at Heathrow Airport, I was so happy to be out of Africa. My wife had been staying in England with relatives and was there to greet me, a doubly happy event.

This was the year of the Summer Olympics in Canada, being held mostly in Montreal with, however, the sailing Olympics competitions being conducted off Kingston. Several small hydrographic surveys were completed prior the commencement of the games, at the request of Games officials. In addition, we provided other vessel support as necessary for the main functions. I was invited to take part in one of these affairs held on board our *Bayfield*, and was glad to help out in making the Sailing Olympics a resounding success. However, I was not impressed by some of the Canadian officials I met, a proper bunch of upper-class twerps. The Sailing Olympics were dominated by moneyed people, certainly on the Canadian side. The spirit and democratic ideals of the Olympics were not obvious to me in Kingston. Perhaps I am being a trifle unfair, but these feelings persist after more than a quarter of a century.

Chapter Thirty-Two

Actions at Home and Abroad

My agenda was fuller than ever in 1977. I was beginning to feel that I was over extended. After a flurry of policy meetings in the department, I was off to Ghana for the third CASLE General Assembly. After my experience in Nigeria, I was not really enthusiastic about returning to Africa, but as a newly elected member of the CASLE executive, I felt that I should make the effort to be there.

As previously, my route was via the UK, and so I was able to touch base with our eldest daughter, who was working in London, and other family and friends, and to meet with colleagues in the hydrographic profession before gathering with other CASLE worthies at Gatwick Airport for the flight to Accra. The aircraft was British and manned by a cheerful crew of English and Scottish lads and lassies, which cheered us up considerably. Among the CASLE delegation on board were Bill and Marg Willey-Dodd from Calgary, Bob Steel, John Bloomfield, and a number of people I had met in Trinidad, such as Randolph Choo Shee Nam of Guyana. The flight was uneventful, but it did give us a wonderful view of the Sahara Desert and an appreciation of its enormous extent.

We were welcomed in Accra by members of the local branch of CASLE, who transported us to a local hotel. We were assured that it was the best in the city, but surely it left a great deal to be desired. It was owned by the government, as was almost everything else in Ghana, a country that had gone downhill since becoming independent. Several coups had occurred, and the country was presently being run by an army cabal. We settled into our ill-equipped hotel and hoped for the best. But the staff was poorly trained, surly, and unable to provide us with items such as cold drinks or even bath mats in the rooms. It was not a very good beginning.

However, one of our delegation, John Hollwey of the UK, had been in Ghana back in the days when it was known as the Gold Coast, and he organized an outing for our benefit the following day, a Sunday—a bus tour out of the city to the botanical gardens. The gardens were famous back in colonial days for the work being carried out on tropical plant development, particularly in the work related to the cocoa plant, the most important part of the local economy. The botanical gardens were still lovely, but signs of decay were everywhere, as a vastly reduced staff tried to cope with the present. It was obvious that no new developments would be taking place here for some time to come. Then back to our lousy hotel along half-finished grand avenues festooned with banners proclaiming the Socialistic Republic of Ghana and all it glories.

We met with the rulers of Ghana the following day in the Assembly Buildings, built back in colonial days. It was a fairly stiff and formal affair, with these military men dominating the proceedings. We were then left on our own to conduct CASLE business, the first matter of importance being the installation of our new president, John Bloomfield of Jamaica. He was a handsome fellow, a graduate of Glasgow University and a Quantity Surveyor. He had all the outward trimmings of a leader, and we all expected great things from him.

Meanwhile, back at the hotel, things were no better—indeed they showed some signs of deteriorating altogether. Linen did not return from the laundry, food was awful, cold drinks were in short supply, electric power shut down, and therefore fans and air conditioning were no longer available. In the middle of all this misery, the management overnight announced a doubling of all prices henceforth. What a mess it all was—we all longed to be somewhere else.

There was poverty on the streets of Accra, and beggars and cripples were a common sight, but much of it seemed self-inflicted by an incompetent government obsessed with left-wing ideas and now headed up by a military dictatorship. The local men, who enlisted in the King's Own West African Rifles and returned from the fighting in Burma against the Japanese full of enthusiasm for a new world, were now wondering what the hell went wrong.

In all fairness, many of our Ghanaian hosts did their best to make us feel welcome. Official affairs were followed by private parties, where we were given a true African welcome. Even the official welcome party was a colourful, wild, noisy affair, with a West African orchestra belting out music that just made your feet itch to get out on the dance floor. Needless to say, after several strong drinks and the drumming of the band, I was out there joining in a celebration of the festivities. I even invited the wife of one of the military junta to join me, which she did to great cheers from the assembly. It was a mad occasion, which seemed to please our hosts. Indeed, the following morning at breakfast, I was saluted by the dining room staff. Judging by my throbbing hangover and that salutation from the hotel workers, I must have put on quite a performance on the dance floor.

There were more serious matters discussed and determined by CASLE during our stay in Ghana. The third General Assembly was able to determine a set of principles and priorities that were laid out in a document known as The Accra Declaration. A number of priorities were underlined for action over the next five years that would underline the enhancement of the profession of surveying in Commonwealth countries, particularly those not now well endowed with these skills. CASLE dedicated itself to fostering closer collaboration between the professions of surveying and land economy and governments, universities, and other professions in the planning and implementation of national development programmes. These goals were to be mainly achieved through professional organization, education, and training. These were all objectives that I believed would be supported by the Canadian Institute of Surveying.

We were given a tour of the port facility and several other government establishments. The government facilities showed signs of decay, and the port, although much funding had been received from the United Nations, also showed similar signs of disrepair. A depressing sight! I picked up a bug of some sort and was dreadfully ill on our last full day in the country. It was not with heavy heart that I boarded our aircraft to fly home. Oh, blessed cold drinks and pretty English hostesses, we were almost home.

As previously mentioned, my daughter Ellen was working in London, and I therefore had the opportunity to spend a little time with her. She was working in the fine arts business and sharing a tiny apartment with three other girls. I think that she was glad to have a chance at a change of pace and to be escorted by her father into relatively expensive restaurants, where she could eat and drink all she wanted. Judging by the knowing looks that I got from waiters and others, I am sure that they thought that I was a dirty old man attempting to seduce an innocent young lady, particularly when I would ply her with a superb white burgundy rated extremely high on the list by wine connoisseurs. We both enjoyed the attention.

There was also a night at the Cleopatra, an exotic dining and dancing establishment that specialized in Greek food and belly dancers. Ellen and Maggie, a younger relative of mine, monopolized the dance floor in a plate-smashing crescendo that cost me a fortune in breakage fees. Meanwhile, I felt a bit like an Egyptian pasha, as a svelte young belly dancer entwined herself around me while trying to extract pound notes from my jacket pockets. Finally, a riotous cat fight erupted between some Brazilian women, which caused us to bid farewell to the Cleopatra.

Back home, a very successful hydrographic conference was held in Burlington. This was our sixteenth annual conference held under the joint auspices of the Canadian Hydrographic Service and the Canadian Hydrographers Association. The site of the conference changed each year, usually Ottawa, Dartmouth, Victoria, or Burlington. In later years, as the CHA and the CHS expanded into new bases, the conference would move to Calgary, Vancouver, and Rimouski, and eventually it become a biennial affair, in lockstep with our US counterparts.

The 1977 conference was held partly in the Canada Centre for Inland Waters and partly in the local Holiday Inn. It was well attended, with speakers arriving from overseas, the United States, and Canada. Although the speakers were largely from government, there were also a number of speakers representing both education and the private sector. As Director of Marine Sciences at CCIW, I welcomed the participants on behalf of the CCIW management committee. The Dominion Hydrographer, Gerry Ewing, and Blair Seaborn, the deputy minister for Fisheries and the Environment, presented opening remarks, and then, under the management of the Regional Hydrographer Central, the work of the conference began.

There were various papers on the developments of technology and technique, but the most interesting papers for me were the Alan Ingham paper on internationalizing hydrographic surveying education at the North East London Polytechnic in England, and a paper by Tony O'Connor, CHS Pacific Region, called "The Chlorine Car Caper," on the search in the Gulf of Georgia for four railroad tank cars, each containing liquid chlorine. The search was a joint effort employing the search resources of both government and the private sector. My old shipmate Dick LeLievre from *Willie J* also presented a paper on the use of Landsat imagery to locate uncharted coastal features.

Rear Admiral Bob Munson, my FIG buddy, was there representing Admiral Powell and the National Ocean Survey of the US, and another old friend, Rear Admiral David Haslam, the Hydrographer of the Navy, UK, presented an eloquent keynote address on the history of hydrographic surveying and the present challenge brought about by such factors as the Law of the Sea, the search for and the exploitation of hydrocarbon resources in the ocean, the deep draught vessels now arriving in numbers on ocean trading routes, and the explosion in new techniques and technology at the service of the hydrographic surveyor, all contributing to great demands on the nautical charting agencies of the world.

Figure 70 - Left to Right: Rear Adm. David Haslam RN, Mike Bolton, Regional Hydrographer Pacific, accepting "Fickle Finger of Fate" from Adam Kerr, Regional Hydrographer Central, 1977

Willie Rapatz, the national president of CHA, presented the Fickle Finger of Fate award to Mike Bolton, who would host the next conference in Victoria. The tradition continued! Meanwhile, all present had a good time, listening to the guitar and fiddle playing by George MacDonald, Ray Chapeskie, and others at the very popular beer seminar, and dancing the night away at the dinner dance.

Figure 71 - Willie Rapatz gives Lighthouse award to Sandy Sandilands, 1977

Figure 72 - A brace of Admirals (Ritchie and Haslam), 1977

While these conferences were very effective, their impact was restricted largely to the English-speaking world. There was a need to expand such technical conferences to a broader audience. The International Hydrographic Organization held conferences once every five years in Monaco, but these meetings were restricted to governments and dealt largely with policy. There was a need for another forum that would bring industry, academia, and government together. Rear Admiral Ritchie of the UK had shown us the way by his work on establishing Commission Four in FIG. We had to find a way to broaden our approach internationally in attempting to solve the problems facing us today and in the future. It was at this conference in Burlington in 1977 that thinking first developed as to how such a conference might be conceived. Such thinking culminated in a decision to hold the First International Hydrographic Technical Conference in Ottawa in 1979, jointly sponsored by the hydrographic commission of FIG, the Canadian Hydrographic Service, the Canadian Institute of Surveying, and the Canadian Hydrographers Association.

The last big event of the year was the FIG Congress in Stockholm, hosted by the Swedish Association of Land Surveyors and the Swedish Association of Real Estate Appraisers. The FIG Bureau was led by President Carl-Olof Ternryd of Sweden and supported by VP William Overstreet of the USA and VP Herbert Mathias of Switzerland. The Congress was honoured by the presence of King Carl XVI Gustaf, who opened the Congress and the exhibition. There were in excess of fifteen hundred delegates and their family members in attendance and coming from more than fifty-eight countries. The nine commissions of FIG discussed two hundred and eighteen invited papers and forty-eight personal papers during their sessions. Simultaneous translation was provided in English, French, and German.

Sixty percent of the papers were in English, twenty-eight percent in German, and twelve percent in French.

The Swedes had arranged many technical tours, including a visit to the Swedish Hydrographic Service headquarters south of Stockholm in Norrkoping. There were also a number of social functions, one in particular in the impressive Stockholm City Hall, which had been built to resemble a baronial hall in the middle ages. The food was great and the liquor of good quality, but the multitude descended upon the feast like gannets, and those of gentler disposition were elbowed to one side in the rush for sustenance. I was accompanied to the event by Bob Munson and his wife Loretta, who shared my view of these starving delegates from Central Europe. We agreed that we would be better prepared for the next major function.

Stockholm was a truly wonderful city, particularly its old town, with wine bars and musicians on every corner. I was transfixed by a harpist playing in one subterranean dive while I quaffed pints of passable red wine. She played beautiful music, seemingly just for me. She was Finnish and obviously bemused by my adoring attention. Members of the Canadian delegation rescued me from my passion and deposited me back in my hotel. A boat tour around the beautiful Stockholm archipelago was a fitting end to our social activities in Sweden.

Commission Four had prepared a programme of seventeen papers for discussion, together with a joint session with Commission Three, where Rear Admiral Ritchie, president of the Directing Committee of the International Hydrographic Bureau, set the stage by delivering an excellent paper on international cooperation in hydrography. The other papers came from the UK, Canada, US, the Netherlands, Germany, France, and Japan. Probably the most interesting papers were the Japanese submission by Uchino and Nagatani on their progress in automated data acquisition and processing, and the submission by John Riemersma of the Netherlands on the use of underwater acoustics for the positioning of offshore structures. There were also reports of the Working Group leaders—Munson of the USA on positioning systems, and Bryant of Canada on data acquisition and processing systems—which gave the commission a clear picture of the state of the art world wide.

Among the other achievements of the hydrographic commission were the recognition that the joint FIG/IHO Working Group on the standards of competence of hydrographic surveyors had completely and efficiently accomplished its task, and that an advisory board on the training of hydrographic surveyors be established to evaluate the adequacy of syllabi and programmes submitted by surveying educational establishments as measured against the FIG/IHO standards. The board would have six members divided equally between the IHO and FIG. The board would meet once each year and report to each permanent committee and Congress of FIG.

The working group reports on positioning systems and data acquisition and handling systems were recognized as being excellent guides, and the work was to be continued, while reporting to the proposed IHTC in Ottawa in 1979 and the next FIG Congress scheduled for Switzerland in 1981.

A new working group was established noting the importance to hydrography of research into anomalies in depths that may not be detected by regular surveys. The working group was charged with making a study of all methods (mechanical, magnetic, acoustic, etc.) and of all instruments that may be used in the detection of underwater obstructions, natural or artificial. The working group would report to the next FIG Congress.

Minuro Nagatani of Japan was elected vice chairman of Commission Four and Admiral Robert Munson of the US secretary for the period of my chairmanship from 1979 to 1981.

The work of the Permanent Committee dealt with a number of items of interest to Canada and the Canadian Institute of Surveying. The Permanent Committee observed a moment of silence in memory of Hans Klinkenberg, a former chairman of Commission Two and a deeply respected member of the Canadian surveying community. Charlie Weir, a former president of CIS, was elected vice chairman of Commission Five. The Appraisal Institute of Canada was approved as a new member of FIG. The next Congress would be held in Montreux, Switzerland, in 1981, with Toronto, Canada, being selected as the Congress venue for 1986. Permanent Committee meetings were approved for Edinburgh in 1980 and Kuala Lumpur in 1982.

Business completed, the delegates departed for home, well satisfied with the work accomplished and the stewardship of the Swedes. I was a little bit sad to leave, as I had a wonderful penthouse room in my hotel overlooking the full sweep of the harbour of Stockholm. My taxi driver to the airport, a buxom, attractive blonde woman, summed up my feelings exactly when she said, "You have enjoyed Stockholm, yah?"

Back home, it was fun to be with my family again, but the children were growing up fast. Indeed, our eldest son Andrew was a young man already, and his siblings were fast catching up. It would not be too long before they had all left the nest.

My wife deserves much credit for keeping things going during my frequent absences. Thank goodness she is a sports enthusiast, who joined wholeheartedly in support of her brood's activities in football and soccer. She was a dedicated fan when her sons were on the field, even loudly berating them if they missed an assignment, or complaining about some idiot who lay prone on the field after a play that had gone wrong—not realizing that the prone figure was one of her boys. After all, it was a night game and there was much confusion! The girls, being girls, did their own thing, Ellen in the UK and Sarah becoming a fair ballet dancer.

That October, I finally had a chance to help some of the CASLE family, by arranging through CIDA for funds to bring four surveyors from the Caribbean to Canada to attend the Third Canadian Colloquium on Survey Education being held at the Manoir du Lac Delage, Quebec. The colloquium was jointly sponsored by CIS, the CCLS, the OQLS, and the Canadian Association of Survey Technicians & Technologists.

Representing CASLE were Harry Armstrong from Jamaica, Mike Griffiths from Barbados, Aldwyn Philip from Trinidad, Godfrey Greenidge from Guyana, and myself. There were about seventy surveyors in attendance, including Derek Cooper, my technical assistant. The colloquium was opened by Dave Usher, the president of CIS. Various speakers followed, including J.Poulin, the deputy minister of the Quebec Lands and Forests ministry. He was followed by Ray Moore, the director general of Surveys & Mapping, Government of Canada. There two papers sparked a great deal of discussion, but it wasn't until the following day that the meeting got going after a session entitled "The Integrated Professional Surveyor" and a panel discussion on the same subject. Herewith follows a summary of my comments as a member of the panel.

> "The papers were all about the land—with the title of the session suggesting a computerized robot. Why not call him 'The Compleat Surveyor' as originally suggested when the colloquium was first planned? Such a surveyor is what we require: a leader in his profession, a catalyst for new ideas and thrusts, a planner for the next decade."

The speakers had completely ignored hydrography. As a member of the panel I forcefully pointed out that we had been in the forefront of new developments for at least the past fifteen years. Our motto could well be "Sextants to Integrated Systems in a Decade."

"Hydrographers have mastered the new positioning systems in the marketplace and developed their own in response to their own specific needs. We are developing side scan sonar mapping and aerial hydrography and are pioneers in resource charting and in many applications of modern techniques to delineate the bottom in ice infested regions."

"How was this accomplished? By sending the best of our surveyors to university for specialized education, and by retraining our technologists not just to operate the systems, but to evaluate the results and modify as required."

"The profession of surveying—if restricted to the Canadian definition of land surveying—is dull and rather uninspiring. No wonder it has had trouble in being recognized as a true profession. The profession must be glamorized in the eyes of the public—the public must see the surveyor as a sex symbol, a chosen one! In the challenge of the offshore, you have that impact, and that is also where you have the truly integrated surveyor. Since 1966 we have added twenty-five percent to the landmass of Canada by international agreement, Law of the Sea negotiations, and by unilateral action—for example, the offshore areas out to two hundred miles off our shores. No nation in the world, not even the Soviet Union, faces such a huge task. Much of the future wealth of this nation lies in that offshore—our survey education system must be structured to ensure that the surveyor is the leader in exploration, development, and production."

"The truly integrated surveyor at sea must not only thoroughly understand the basics of surveying and the flexibility and limitations of the new technology, but he must also have a broad knowledge of other scientific disciplines—geophysics, oceanography, geology, biology, and meteorology. He is the owner when dealing with the ship's Captain. Ideally he should be a mariner himself. Who else understands that hostile environment so well?"

"If we require a model of the Compleat Surveyor—look no further than James Cook. His range of interest and expertise encompassed astronomy, botany, biology, hydrography, navigation, seamanship, and of course, anthropology. His sojourns in the Tahiti and elsewhere in the South Pacific, his obvious delight in the Tahitians, and his appreciation of their charm and beauty made him very, very human, compleat in every way."

I had perhaps been carried away a little, as I am apt to do at times. But I was angry at all these leading lights in the surveying and surveying education world who were so firmly anchored to the land. Other sessions followed on the Profession/Academic interface, an excellent presentation by John Mclaughlin of the University of New Brunswick and another in a similar vein by David Lambden of the University of Toronto. All in all, it had been a worthwhile exercise, which had appealed to the contingent from the Caribbean as well as those of us from across the country.

Adam Kerr departed the region in late 1977 to take educational leave in the UK, where he eventually obtained a Masters in Marine Law from the University of Wales. I was sorry to see him go. He had been a burr under my saddle at times, but I never doubted his integrity and enthusiasm in the cause of the Canadian Hydrographic Service.

Chapter Thirty-Three

Major Changes in the Hierarchy

Changes occurred over a period running from mid-summer 1977 to later in 1978 that had some significance to me personally. Art Collin departed to become Assistant Deputy Minister Atmospheric Environment, the department of Fisheries and Oceans was formed, and Gerry Ewing became my immediate superior as Assistant Deputy Minister of Ocean and Aquatic Sciences. My organization in Burlington was renamed the Bayfield Laboratory for Marine Sciences and Surveys in recognition of its expanded role in the conduct of physical oceanographic studies in the High Arctic and the St. Lawrence River and its approaches. I therefore became the Acting Regional Director General of the Bayfield Laboratory. Also in 1978, Ross Douglas of Atlantic Region joined us as the new Regional Hydrographer Central.

In addition to my administrative policy development duties, I was still active in a number of professional associations with interests close or peripheral to ships and seagoing trade. In early 1978 I officiated as president of the Marine Club at their annual dinner held at the Royal York Hotel. This was the culmination of a week of meetings between various sectors of the marine industry operating on the Great Lakes, on matters concerning ship operation and the flow of cargoes. As usual the place was packed with about sixteen hundred attendees, and the head table was a glittering display of the movers and shakers of the industry. This year the keynote speaker was the head of the Seaman's International Union, a roughneck from the waterfront, whose presence there was not enthusiastically endorsed by all. He had a bodyguard who accompanied him wherever he went, making things awkward during the course of smaller social functions. His speech was delivered well, but caused some frowns on the part of the titans of the shipping world. I do not remember how our organizing committee arrived at a decision to invite the SIU, but I felt that it was a mistake and was glad when the affair was concluded.

There was a hydrographic conference in Victoria that was a must to attend, it being our former home for many years. It was grand fun and featured a cornucopia of first-class presentations on the state of the art in hydrographic surveying. It also was memorable for its social functions, where a strolling troubadour followed us around town from pub to pub strumming away as we sang ditties and songs of the sea. Among the foreign guest speakers was the head of the Soviet Naval Hydrographic Service, who was quite a character and difficult to control at times. His name was Chandabylov and he loved Scotch whisky and the ladies, not necessarily in that order. He invited a number of women associated with the conference up to his room in the Empress Hotel—an invitation that could be fraught with all sorts of implications. He asked my wife to buy a number of small items for him, baby items for his grandchildren, and wanted her to deliver them in person. As his reputation was well known, she graciously declined and had them delivered to him by courier. He was an interesting man to talk to. While obviously a political animal, he had had an interesting life, being at one time during WW II Russian Naval Attaché to the Royal Navy based in Scapa Floe. He knew Scotland well and could produce a respectable Scottish accent. He was quite emotional and kissed us goodbye outside the Empress Hotel, leaving our ADM to ensure his safe passage to the Soviet Embassy in Ottawa.

CASLE had organized a regional meeting in Barbados, where I had been invited to present a paper. Naturally, I took my better half along for the event, and I believe that she thoroughly enjoyed herself. She must have, because late at night when the pool lights were dimmed and her third planter's punch was having its effect, she decided to go skinny-dipping. Next morning, most of the CASLE delegates seemed to know about her daring, and one man swore that he saw her from his balcony, exiting from the pool like Lady Godiva. And so she achieved CASLE fame.

My paper was on the Law of the Sea and its implications for the hydrographic surveyor, with particular emphasis on the Caribbean region. It was well received by the relatively small audience that attended the meetings in Bridgetown. Mike Griffiths, of the Association of Land Surveyors of Barbados, made sure that we were well looked after and saw many of the delights of the island. We particularly enjoyed watching the flying fish off the southeast coast and watching the young lads and lassies swimming like fish themselves in the clear blue water.

The event of the year was a visit to Paris to attend the Permanent Committee meeting of FIG. Because FIG had held its inaugural in Paris in 1878, it was a meeting of great significance. I had never been to Paris before, and so found the city a most attractive place. As an amateur historian, I just lapped up the sense of history and pageantry displayed all around me. As a gourmand aspiring to be thought a gourmet, I relished the food and the wine. I was in seventh heaven!

Unfortunately, I had to work also, but tried to balance my longings with my duty. The main work of the Permanent Committee went ahead as usual in all three official languages of FIG: English, French, and German. Naturally, French got a big play as everyone celebrated the centennial.

Additionally, we were wined and dined at a special function at the Paris Opera House, and were able to take advantage of the opportunity to visit Notre Dame Cathedral and take a romantic trip on the River Seine. There was also time for an evening visit to Montmartre and the nightlife along the left bank of the river. My final treat was a visit to Sacré Coeur Basilica, a magnificent edifice that sits on the summit of a hill dominating the city. I had always had mixed feelings about the French nation, but the sights of Paris partially converted me. Paris is well worth a mass, as a French king once proclaimed.

Commission Four business was rather restricted, as we were only three officers, Chairman Marc Eryres, Secretary Gerry White, and myself as Vice Chairman. As I had the planning for the International Hydrographic Technical Conference, Ottawa 1979, and the FIG Congress, Montreux 1981, well in hand, it did not matter too much. However, I was needed for the Permanent Committee deliberations on future venues for the FIG Congress and other matters of concern. The vote at the Permanent Committee meeting in Ibadan had narrowly gone against us, and because Bulgaria would host the 1983 Congress, we pushed vigorously for a firm decision on the 1986 venue. The Permanent Committee agreed that it was Canada's turn in Toronto in 1986. Charlie Weir, CIS, and vice chairman of Commission Five, joined me in pressing our case. We could see a lot of hard work stretching ahead of us for the next several years.

I spent some time with Bob Munson in Norfolk, Virginia, where he had taken up an appointment as Rear Admiral and commanding officer of the NOS base Atlantic Marine Centre. As he was now secretary of Commission Four of FIG, it was necessary that we liaise on FIG matters, particularly the coming Ottawa IHTC in 1979 and, of course, the 1981 FIG Congress scheduled for Montreux, Switzerland. It was also a grand opportunity to get a feel for the US way of doing things. Their base in Norfolk was quite large, with docking facilities for several large and medium-sized vessels, together with ship support facilities and laboratories devoted to chart compilation and photogrammetric interpretation.

NOS, while a civilian arm of the US Department of Commerce, nevertheless operated in quasi-military style, with officers holding the equivalent of naval ranks. However, the ship's crews and other support staff ashore were civilian. The key unit was the controlling Officer Corps. These officers all held university degrees in various specializations. These qualifications were mostly concentrated in the engineering and oceanographic disciplines, and their training in seamanship and navigation seemed to be much of an ad hoc approach on the job. They were very bright, but did not give off that unmistakable aura that defines the seaman. However, their end product, the nautical chart, was of high quality. As most Canadian hydrographers had little background in seamanship and navigation, we were hardly an example that the NOS might be persuaded to follow.

Bob and his good lady Loretta made me most welcome in Norfolk and became our very good friends. We also had a link through Rotary membership, which gives me a reason to comment briefly on my accession to the presidency of the Rotary Club of Burlington, Ontario. I enjoyed being a Rotarian mainly because the membership of Rotary comes from such a wide and varied slice of society. I was meeting people I would never encounter while primarily pursuing my hydrographic profession or the associated world of ships and the sea. I worked hard for several years for my club, but was never completely sold on the rah-rah fund-raising effort required to raise funds for charitable purposes. I would much rather donate cash to such ventures than give of my time, which I considered absolutely precious. However, I believe that I made a respectable president and have remained a good Rotarian to the present day.

Chapter Thirty-Four

Triumphs – Rewards – Reflections

The time has come to perhaps think about ending this epistle. In the preceding chapters I have gone from being an almost penniless emigrant from Britain to a comfortable bureaucrat in the service of the Government of Canada, summarized perhaps best as "Navigator to Hydrographer to Bureaucrat." In 1979, I was confirmed Director General of the Bayfield Laboratory of Marine Sciences and Surveys, which was to be the crowning achievement of my career in government.

Figure 73 - The plaque commemorating the first IHTC, Ottawa 1979, sponsored by Fédération Internationale des Géomètres, Canadian Hydrographic Service, Canadian Hydrographers Association, and the Canadian Institute of Surveying

It was the year that the First International Hydrographic Technical Conference became a reality in Ottawa. As the leading promoter of the need for such a conference, and vice chairman of the steering committee that helped unite the efforts of the Fédération Internationale des Géomètres, Canadian Hydrographic Service, Canadian Hydrographers Association, and the Canadian Institute of Surveying to produce the IHTC, I was enormously pleased at its success. The IHTC deserves a chapter of its own to celebrate its truly international flavour and the melding of the private sector, government, and academia in technical and scientific endeavour. It will, however, have to await the next volume of my memoirs. This was also the year that Steve McPhee became the new Dominion Hydrographer.

Meanwhile, my foreign travel continued, first of all to Brno in the heart of Europe, to attend an FIG Permanent Committee meeting. Brno is a small city in Moravia surrounded by gently sloping hills and lush, grass-laden valleys supporting much agricultural activity. These were still the days of the Iron Curtain, and transportation within Czechoslovakia was difficult, particularly after the Russians had just very recently increased their military presence in the country. I must have had my documents examined a score of times before arriving by small aircraft in Brno after a short flight from Prague. Our hotel would have rated two stars in the west, with narrow beds and duvets that kept sliding onto the floor. Food coupons had to be purchased in the hotel for meals, which were of a very mediocre standard. However, the black market was alive and well and controlled in the hotel by the elevator operators. Good food could be had in selected restaurants that would only accept payment in US dollars. We were shadowed everywhere we went by peculiar-looking men in thick overcoats and army boots, and the bars were filled with whores who obviously reported back to the local version of the KGB.

In addition to the members of the Canadian contingent such as Hugh O'Donnell and Charlie Weir, I found myself often in the company of Terry Sudway, the delegate from Ireland. Like me, he had a passion for history, and after discovering that the site of the battle of Austerlitz was close by, we managed to hire a taxi to take us to the battlefield and view the museum and the countryside around. It was an impressive sight—the sweeping vista of grassland where Napoleon had his great victory over the Austrians and the Russians. Sad also to contemplate the carnage of slaughter that must have overhung the battlefield on that fateful day so long ago.

Sudway and I spent a lot of our time together discussing the situation in Northern Ireland. It soon became obvious that he was a strong supporter of the Irish Republican Army and condoned the bombings and slaughter of innocents in the cause of "freedom." As a Scottish Protestant with probably a modicum of Ulster blood in my veins, I was appalled! There may well have been wrongs that needed to be addressed, but the killing of innocent women and children was beyond the pale. We continued to be friends for a number of years, but from that day onward I was wary of his true allegiance. He did have a highly developed sense of humour, which came to the fore when we made the mistake of cheering and waving in support of some soldiers passing by in a convoy of trucks, only to find out from our driver that these were the despised Soviets. Oh, well, you cannot be right all the time.

There was the usual welcoming ceremony for our group, although the local dignitaries all seemed to be Communist party officials who loudly proclaimed the brotherhood of man and emphasized the word "peace" at every opportunity. I regarded them with some suspicion and skepticism. I was all for peace, but not at the price that I suspected they had in mind. I was to find that this simple, beautiful word would be used often by the representatives of the Communist states that had recently become members of FIG.

The work of the Permanent Committee was fairly straightforward, with our job being to emphasize the facts that Canada would be hosting the 1986 Congress in Toronto and that we welcomed one and all, from behind the Iron Curtain or not. My travel home allowed me one day in Prague, where I was able to see a bit of the old city. My hotel was quite modern and comfortable, but their holding of my passport enraged me, as it had in Brno. I regard such action as those of a police state that would leave me unable to properly identify myself in an accident or other trouble encountered away from the hotel. However, I managed to leave Eastern Europe without further upset and, after a short family visit in the UK, returned safely home.

The other overseas travel was to Hong Kong to take part in the CASLE Asia meeting. I had passed through Hong Kong several times previously on my way to other places in Asia, but this meeting presented an excellent chance to see a bit of the dynamic Crown Colony of Hong Kong. While I was taking part in the CASLE meetings on Hong Kong Island, it was still possible for me to travel up to the Peak and see various other sights on the island and in Kowloon and the territories beyond. As a Rotarian, I had access to several Rotary clubs, where I was made very welcome. Although some clubs conducted all their business in English and others in Cantonese, the sincere welcome was the same. The business people hoped that the relationship with mainland China would continue to improve, and other folks just wanted to get on with their own lives. Only once did I hear anyone say that Hong Kong should be returned to China—the travel guide who took us up to the peak. I loved the great variety of Chinese cuisine available everywhere, and made sure that I sampled more than my share.

The CASLE meeting was hosted by the Surveyors of Hong Kong, many of whom were members of the Royal Institution of Chartered Surveyors in the United Kingdom. We were entertained by members in their homes, some of which were quite palatial and high up the peak, with magnificent views across Hong Kong Harbour. Obviously, some expatriates were doing exceedingly well in Hong Kong. The main items of business at the CASLE meetings were to report on the CASLE general assembly in Accra, prepare for a meeting the following year that would take place in Papua New Guinea, and plan for the next general assembly, which was to be held in Ottawa. All of that kept me quite busy, but I was still able to find time to wander through the bazaars and alleyways in search of curios of all sorts. For a collector like me, it was a wonderful experience.

My flight to Hong Kong had been by Pan Am from San Francisco, and I was able to spend some time with John Bloomfield, the CASLE president, who had just left his home in Jamaica. He was an interesting man, born in Jamaica with a final education in Scotland, and married to a Scottish woman. He was handsome, with considerable drive and enthusiasm, and had a strong, sonorous voice, which would have served him well in a pulpit. We fortified ourselves with the occasional alcoholic beverage en route across the Pacific, and I was served the largest Drambuie I have ever had by a stewardess who was not aware of its potency. I was duly thankful and toasted her generosity as she collapsed in a nearby seat to rest her aching feet. She was quite elderly for a stewardess, only on board because her union insisted. I hope she duly recovered and flew many another mile while chatting away about her grandchildren.

My flight home was dull by comparison, and I was back in Burlington without further ado.

As mentioned earlier, the name of my establishment had become the Bayfield Laboratory for Marine Sciences and Surveys in 1978, and I was confirmed as Director General in 1979. The Bayfield Laboratory was quartered at the Canada Centre for Inland Waters in Burlington, close to the heart of Canadian industrial power on the Great Lakes of North America. The name "Bayfield" had some significance here, as the first real hydrographic surveys of the Great Lakes were conducted largely under the command of Lieutenant Bayfield RN shortly after the end of the war of 1812 and the Napoleonic wars. Indeed, after completing the nautical charting of the Great Lakes, he went on to complete the surveying of the St. Lawrence River and Gulf before retiring in 1856.

Henry Wolsey Bayfield was born in Hull, England, in 1795, and joined HMS *Pompey* as a supernumerary before he had reached his eleventh birthday. He served at sea during the Napoleonic wars, being stationed latterly on HMS *Beagle* operating out of Halifax and Quebec City. His initial hydrographic surveys commenced in 1816 and were conducted with a minimum of support, being denied a larger vessel and sufficient crew. However, things did improve, and by 1828 he had completed the Great Lakes charting and went on to commence a detailed survey of the St. Lawrence River and Gulf while basing himself in Quebec City.

His interests went beyond nautical charting: collecting geological specimens, studying the phenomena of weather such as fog, the tidal currents, the aurora borealis, terrestrial refraction, and the intricacies of the ship's chronometer. His was certainly an enquiring mind.

In 1841 he transferred his headquarters to Charlottetown, Price Edward Island, to enable easier access to the seas around Gulf of St. Lawrence. His final survey was that of Halifax Harbour and surroundings, which he completed 1853. By now he was slowing down, and in 1856 he retired with the well-earned rank of Rear Admiral. He was a most gifted and zealous surveyor.

He is honoured ashore in villages bearing his name in Ontario, New Brunswick, Nova Scotia, and Prince Edward Island, and in the naming of a series of hydrographic vessels *Bayfield*, in recognition of his contribution to the opening up of Canada. It is therefore entirely appropriate that the laboratory in Burlington, Ontario, be named after such a man. I am proud to be associated in a very small way with his name.

Figure 74 - Rear Admiral Henry Wolsey Bayfield RN, 1795–1885

I now have time to reflect on many things: what I believe I did well and what I know I could have done better or perhaps differently, my weaknesses and my strengths, my luck at times, and my strongly held views, which at times led me into confrontations that could probably have been avoided. So to put it all into context, I will comment on the world and national scene that I recall from 1948 to 1979, and then deal with the more personal matters.

In 1948 the Soviet blockade of Berlin was clear evidence that we faced difficult years ahead. This event was followed closely by the victory of the Communists in China and shortly afterwards by the invasion of South Korea and a full-scale Korean War involving the US and the United Nations against North Korea and China. The next confrontation surfaced when Cuba and Russia combined forces to place missiles and nuclear warheads in positions threatening the mainland of the United States. This was a severe crisis that threatened all mankind. Meanwhile, insurgency in colonial possessions in South East Asia eventually escalated into a full-scale war in Viet Nam, while the establishment of the state of Israel brought war to the Middle East, and by 1979 the price of oil was having a serious effect on the economy of the globe. It was a dangerous world.

Nationally, we finally turfed out the Liberals and John Diefenbaker became our Conservative prime minister. He frittered away his government's chances to retain power by misreading the "Cuban Crisis" and upsetting our American ally. Under Pearson, we got a new flag that was supported by one third of the population, opposed by another one third, and ignored by the remainder. Democracy at work? My children and grandchildren love it!

We then entered the Trudeau era, he who proceeded to ruin the country by his fiscal policies, politicizing the higher ranks of the civil service and weakening our armed services, and making overtures to the Communist states while ignoring his long-term allies. He ignored his own cabinet and used the PMO to enforce his decisions directly on departments. In 1979, the electorate gave the Conservative leader Joe Clark a minority government. However, Joe could not count properly, and we were faced with another spell of Trudeau. Regardless of the pall that such a fate exerted over a large part of the population, Canada was still a good country!

Provincially, my experience was limited to Ontario and British Columbia. In Ontario we seemed to have had the Conservatives in power forever, providing solid if dull government. BC was much more volatile, with a right/left bias that, while troublesome at times, ensured that life was never dull. Certainly, having a premier such as W.A.C. Bennett, who led a right-wing government but who had no hesitation about taking over responsibility for power distribution and an expanding ferry system, was anything but run of the mill. There was also, of course, Tommy Douglas from Saskatchewan, who brought medicare for all to his province and thereby influenced the Government of Canada in adopting a similar programme nationally.

In reviewing my own personal experiences, I find myself having to face up to my own weaknesses and prejudices, as well as to comment upon what I feel were worthwhile achievements, to recognize where Lady Luck played a prominent role and where I benefited from selfless help of others.

Where do I start?

My weaknesses are many; over indulgence in good food and liquor have been my downfall since my early days at sea. The liquor part of it sometimes led to indiscretions on my part, sometimes in speech, and from time to time emboldening me enough to respond to the provocation of a pretty face or a shapely figure. I have now grown older and perhaps wiser, but in 1979 I was still capable of being a silly ass! My other big weakness did not need liquor to bring it to the surface: my temper became very occasionally a weapon that could destroy opponents or stir them up to retaliate with interest. I do not believe that I deployed my temper against subordinates, but I surely took on my equals and several of my seniors in rank. It was an impediment to my further advancement. Additionally, I always hated attention to detail, leaving subordinates to dot the i's and cross the t's, while I concentrated on the broader picture. Upon reflection, I know now that closer scrutiny to detail might have served me well.

Under the rules regarding language training that went into effect in 1967, I was grandfathered and did not have qualify as being proficient in the French language. When I became Regional Hydrographer, I had the opportunity to volunteer to take such training, but was up to my hips in learning the intricacies of managing my region and so declined. I would like to have taken French language training at an appropriate time, to enhance my understanding in a situation where about twenty percent of my staff were French Canadian. I never did find the time to get away from the rat race.

I have always admired people who can "think on their feet," responding well to pointed hostile questions. Somehow or other, particularly in a tense environment, my mouth and my brain refuse to act in tandem, and I find myself mouthing a different response from that being urged by my brain. Excellent in preparation, but weak in response!

My final weakness to confess is my attention to my job and my profession, sometimes at the expense of my family. Now that my children are all grown up with families of their own, I know that I missed part of their growth into adulthood and should have been there to help my wife, particularly in matters relating to their education. My wife did a sterling job on her own, under difficult circumstances, but I should have been there more to help. Thank goodness they all turned out to be reasonably decent citizens of this country. It is only now that I realize how lonely she must have been, without any members of her own family living close by in support.

My strengths were in persuasion, by both the written word and in small group meetings. I could write well, as my very many papers attested, with regard to technical and policy content. I was persistent, indeed at times perhaps obsessively so. I had set myself clear goals and objectives and worked hard to attain them. I liked and admired most members of my staff and believed that these feelings were reciprocated. I was decisive when required and a reasonably good negotiator when a certain amount of finesse was necessary.

What therefore went right for me? I made the correct first move when we came to Canada in 1948. It had been a bit of a roller coaster in the early years, but I was now a senior civil servant with all the power and fluff that such a position denotes. I had enjoyed my time in the Arctic and was lucky enough to take part in the *Richardson* drama in the ice off Point Barrow. I am sure that my Arctic endeavours and my prominence in the push for a training policy in the Canadian Hydrographic Service were factors in my selection as Regional Hydrographer. Other promotions were, I believe, associated with expanding the region's activities in a period of dramatic change in departmental responsibilities. Lady Luck also played her part.

What went wrong or was not achieved? I had strongly advocated a continuing mariner strength within the ranks of the CHS, to ensure that we maintained an ongoing interface with the end users of the nautical chart—the maritime industry. I pointed to people like myself and Adam Kerr, who had served in dual capacity as Master and hydrographer-in-charge, and sought to broaden the concept. It was all in vain, as the CHS training plan had no room for anyone other than university or technological institute graduates. There was one small experiment undertaken on the Atlantic coast, but elsewhere, the mariners died off or were not encouraged. Even after all these years, I feel that it was a great mistake, and so noticeable when we interact with hydrographers from other nations who still have that dual role.

As noted earlier, I had strongly advocated "contracting out" of hydrographic surveys and chart compilation where it made economic sense, but by 1979 in Canada, the effort was still being hamstrung by lack of dedicated funding or long-term commitment on the part of CHS management and staff and by underlying subtle attempts to sabotage such undertakings. The efforts of companies such as McElhanney, Terra, and Comdev Marine to comply with existing CHS standards were commendable, but they found it all an uphill struggle. Today, other countries such as the US, UK, the Netherlands, and Australia, just to mention a few, have broadened the scope of the private sector's involvement in hydrography, while we in Canada are no longer a leader in such activities. Indeed, newly developed countries such as Malaysia are showing the way. Another missed opportunity!

I find that as I grow older I become more cranky and dismissive of modern trends. The politically correct infuriate me. As one who was called in my seafaring days by my Sassenach shipmates a "porridge eater," a "tartan toe rag," and a "web-footed highlander," among many other insults of an unprintable nature, I was the object of many politically incorrect jibes. I took it all in my stride, knowing full well that I was of better breed than those who hurled these insults at me. So just forget about the politically correct thing to say and roll with the punches, as we used to in days gone by.

I am also a bit puzzled by feminism and its apparent aims. In my first book *Mandalay to Norseman*, I mentioned my bemusement at seeing a Russian heavy-lift vessel in Murmansk with an entire female crew, including the captain and other officers. Since then we have seen women advance in every sector of society and make their voices heard. By 1979 I had three young women leading divisions reporting to me. Now most of the students entering university are female, and men appear unsure of how to handle the changing landscape. To feminists, I say back off a bit and think about what you might be doing in furthering your aims at the expense of the male half of society.

I hope to continue this dialogue in my third book, which will deal with events from 1979 to the present day.

ISBN 1-41204592-4